Advance praise for Project Planning and Management for Ecological Restoration

"Adding to an already rich series on ecological restoration texts, Island Press delivers this book chock-filled with over 100 years of accumulated knowledge and on-the-ground experiences by three luminaries in the field of ecological restoration. With a focus on project management Rieger, Stanley, and Traynor effectively bring the philosophical, ecological, and social aspects of ecological restoration to a practical reality. Students, practitioners, and, yes, even philosophers stand to benefit from reading this book."
—Nik Lopoukhine, Past Chair of Society for Ecological Restoration

"I have seen many well-meaning restoration projects fail due to poor planning, but that need never happen again. The authors lend over a century of combined experience in ecological restoration design at a variety of scales to describe the detailed analysis and planning process necessary for successful project design, implementation, maintenance, and evaluation."
—Steve Windhager, Executive Director, Santa Barbara Botanic Garden

"This sorely needed book presents a comprehensive view of restoration project management from a biological perspective. It will be especially useful to those now entering the field of ecological restoration as practitioners. I wish I had this book when I began my career as a practicing restorationist."
—Tom Griggs, Senior Restoration Ecologist, River Partners

"Whether you are a restoration practitioner, student or researcher, this is a necessary handbook of guidelines for all aspects of restoration and management. Nonspecialists and ecologists alike can readily take in the complexities of planning, risk management, and goal setting for regenerating ecosystems. Case studies and flow diagrams are presented to bring the world of planning to life."
—Edith B. Allen, Professor, Department of Botany and Plant Sciences and Center for Conservation Biology, University of California, Riverside

About Island Press

Since 1984, the nonprofit organization Island Press has been stimulating, shaping, and communicating ideas that are essential for solving environmental problems worldwide. With more than 800 titles in print and some 40 new releases each year, we are the nation's leading publisher on environmental issues. We identify innovative thinkers and emerging trends in the environmental field. We work with world-renowned experts and authors to develop cross-disciplinary solutions to environmental challenges.

Island Press designs and executes educational campaigns in conjunction with our authors to communicate their critical messages in print, in person, and online using the latest technologies, innovative programs, and the media. Our goal is to reach targeted audiences—scientists, policymakers, environmental advocates, urban planners, the media, and concerned citizens— with information that can be used to create the framework for long-term ecological health and human well-being.

Island Press gratefully acknowledges major support of our work by The Agua Fund, The Andrew W. Mellon Foundation, Betsy & Jesse Fink Foundation, The Bobolink Foundation, The Curtis and Edith Munson Foundation, Forrest C. and Frances H. Lattner Foundation, G.O. Forward Fund of the Saint Paul Foundation, Gordon and Betty Moore Foundation, The Kresge Foundation, The Margaret A. Cargill Foundation, New Mexico Water Initiative, a project of Hanuman Foundation, The Overbrook Foundation, The S.D. Bechtel, Jr. Foundation, The Summit Charitable Foundation, Inc., V. Kann Rasmussen Foundation, The Wallace Alexander Gerbode Foundation, and other generous supporters.

The opinions expressed in this book are those of the author(s) and do not necessarily reflect the views of our supporters.

PROJECT PLANNING AND MANAGEMENT
FOR ECOLOGICAL RESTORATION

Project Planning and Management for Ecological Restoration

John Rieger, John Stanley, and Ray Traynor

ISLANDPRESS

Washington | Covelo | London

Figure 6-6 is adapted from figure 7 in "The Island Dilemma: Lessons of Modern Biogeographic Studies for the Design of Natural Reserves," by Jared Diamond, *Biological Conservation*, vol. 7, no. 2, 1975, pp. 129–46. Reprinted with permission from Elsevier Science.

Library of Congress Cataloging-in-Publication Data

Rieger, John P.
Project planning and management for ecological restoration / John Rieger, John Stanley, and Ray Traynor.
    pages cm. – (The science and practice of ecological restoration)
    Includes bibliographical references and index.
    ISBN 978-1-61091-363-8 (cloth : alk. paper) – ISBN 1-61091-363-9 (cloth : alk. paper) – ISBN 978-1-61091-362-1 (pbk. : alk. paper) – ISBN 1-61091-362-0 (pbk. : alk. paper) 1. Restoration ecology–Planning. 2. Restoration ecology–Management. I. Title.
    QH541.5.R45R54 2014
    577.068–dc23

2013045523

♻ Printed on recycled, acid-free paper

Manufactured in the United States of America
10 9 8 7 6 5 4 3 2 1

Keywords: Island Press, Society for Ecological Restoration, ecological restoration, restoration practitioner, restoration ecology, ecological function, ecosystem, stewardship, restoration site, project planning, project design, project implementation, project management, project budget, project cost, project manager, after-care, restoration strategies, stakeholders, project schedule, Gantt chart, risk management, site analysis, landowner, site improvements, extant reference site, historical reconstruction, remnant patch, design elements, landform modification, water delivery system, soil preparation, erosion control, slope stabilization, seed collection, seed viability, plant propagule, plant translocation, irrigation, bill of materials, site inspection, infrastructure, maintenance, weed management, risk management, weed control, invasive species, exotic pest, monitor, monitoring protocols, monitoring program, Los Peñasquitos Canyon Preserve

CONTENTS

Practitioners of ecological restoration spend their careers cleaning up someone else's environmental messes. It's human nature to exploit natural resources as if they were gifts from the gods, rather than respecting the natural world as if our lives depended on it—which they do. If we want nature to sustain us, we must reciprocate and sustain nature. Our causative role in global warming has brought this lesson home to us—in spades. The time has come to clean up our messes, and this book shows how to do it.

A century ago, conservationists argued that entrepreneurs should practice wise use of natural resources and keep their hands off protected reserves. Now, restoration practitioners find themselves fixing nature that was desecrated by successive generations of entrepreneurs. Those responsible for the damage would rather keep their hands off impaired ecosystems and protect their own cash reserves. But the time for walking away from the ruination of nature has long gone, and ecological restoration is becoming the new modus operandi of how we do business.

The impetus behind many ecological restoration projects consists of government mandates and environmental permitting. However, the public policies that drive ecological restoration are not homogeneous. Instead, they are mirrors for a range of public opinions concerning natural resource management. Environmental regulations echo a sophisticated understanding of our dependence on the natural world, yet they contain loopholes that admit those who seek short-term profits. Other loopholes reflect mandates in the missions of public agencies, which are not always compatible with restoring ecological wholeness and sustainability as advocated by ecological restoration.

Our emerging profession is buffeted by a plethora of conflicting expectations and a lack of consensus on professional standards. This confusion reflects the youthfulness and fluidity of our discipline and the social, ecological, and economic complexities inherent in every project. Restoration practitioners must skillfully navigate these opposing tendencies if they are to attract professionally rewarding contracts and employment that advances their professional and personal aspirations. This book explains the tools and the mind-set that allow practitioners to face these challenges successfully.

Young practitioners, whose exposure to environmental concerns stems primarily from classroom experience, may suffer shock upon graduating into the world of budgets and permitting criteria that unceremoniously dampen environmental fervor. They will be better prepared for the workforce if they absorb the messages of this book during their academic preparation.

Restoration project sites are the locations where idealism and reality clash. Ecological restoration practitioners are caught squarely in the middle, both as participants and as referees. They must know what they are doing, because they have mortgage payments due each month. That's where this book fits in. It was written by experienced practitioners who have waged project-related battles and faced challenges—successfully. They have distilled decades of project experience and are imparting their knowledge for the benefit of less seasoned members of our discipline and for the elucidation of those who are more peripherally engaged in restoration project work.

Authors John Rieger, John Stanley, and Ray Traynor are dedicated to the twin ideals of improving environmental well-being and refining restoration practice. Their overriding concern throughout this book is how practitioners can clean up environmental messes and return ecological integrity in a way that accommodates contemporary realities and constraints yet advances the profession of ecological restoration—not an easy assignment!

If the ecological restoration movement is going to maintain itself and further gain momentum, it will be because of the dedication of competent practitioners. This book builds professional competence with practical advice presented within a coherent framework of action. Its contents generate the confidence that practitioners need if they are to maintain their dedication during the complications that crop up in nearly every project.

Sound project management is a recurring theme throughout this book. Too many restoration projects in the past were predicated on naive idealism, only to suffer from insufficient planning and budgeting, and from inadequate scheduling and coordination. Another theme is the importance of satisfying the sometimes conflicting values of those who are most directly affected by a restoration project: the stakeholders. This is no small concern to practitioners. Stakeholders may become future clients, or at least their opinions could influence who is awarded the contract for the next job.

We can imbue public consciousness with the merits of ecological restoration incrementally. Each project adds to the whole, no matter how difficult the unavoidable challenges were that constrained its outcome. Eventually, we will build an undeniable body of project work, one that more than justifies our efforts, and with it a profession that generates effective standards of practice. Our good work will elevate public awareness of the benefits we receive when we take care of nature. We can't accomplish this grand enterprise in isolation. As professionals, we need to help one another, and that's exactly what John Rieger, John Stanley, and Ray Traynor will have done for the practitioners who read this book.

Andre Clewell
President Emeritus of SER
Ellenton, Florida

Ecological restoration is becoming an increasingly important component of land stewardship throughout the world. After several decades of observing the decline in biodiversity and function of natural areas, land stewards have initiated the process of guiding their properties into healthy, functioning ecosystems through a variety of ecological restoration activities. The process of ecological restoration has no magic formula, as each site is unique, with its own set of degrading causes, ongoing stressors, and operational constraints. The challenge of managing a complex array of stressors and the lack of a routine system of project development are the root causes for the failure or diminished success of the many ecological restoration projects that we have observed. We have been involved in conducting ecological restoration projects for the past thirty years. During this time, we have experienced a wide range of projects with varying causes of degradation as well as several vegetation types. Over the years, we have seen several projects exhibit short-term improvements, only to see the gains quickly reversed by the original damaging source, which was never properly addressed.

Ecological restoration projects are performed by a wide range of individuals and groups with an equally wide range of experiences and knowledge. Today, many professionals and volunteers are involved in the various aspects of ecological restoration project work. Some consider themselves professional practitioners of ecological restoration, others work as professional land managers or stewards, and still others assist restoration project managers by providing stakeholder input, technical expertise, fund-raising, or manual labor.

This book has been designed as a practical means of executing an ecological restoration project using a methodical approach. We provide tools and knowledge to ensure that your ecological restoration efforts will succeed. When undertaken thoughtfully, thoroughly, and objectively, ecological restoration projects are more likely to meet their goals.

In 2000, the Society for Ecological Restoration (SER) published the first edition of *Guidelines for Developing and Managing Ecological Restoration Projects* (Clewell, Rieger, and Monroe 2000) in response to the need to establish norms for the planning and implementation of ecological restoration projects. The senior author of this guidebook participated in the development of these guidelines and is a coauthor of the second edition, posted on SER's website (Clewell, Rieger, and Munro 2005) and reprinted in the first edition of the book *Ecological Restoration: Principles, Values, and Structure of an Emerging Profession* (Clewell and Aronson 2007). Although this guidebook does not completely parallel the steps outlined in the SER's *Guidelines*, it builds on that foundation.

This guidebook will make an excellent companion text to restoration ecology textbooks. It provides the hands-on application of what is typically learned in restoration ecology courses, by providing the practical tools and methods of ecological restoration that are often missing in formal college classes.

Applying the principles and practices described in this book should enable motivated individuals to plan and implement successful restoration projects. It is our hope that improved practice in the field of ecological restoration will translate into a reversal of the degradation of our planet and the development of sustainable livelihoods for peoples who want to reestablish a healthy relationship with their environment.

This book evolved out of feedback received during various restoration workshops taught by John Rieger, John Stanley, and Ray Traynor along with a host of associate instructors. Without the sponsorship of these workshops by the Society for Ecological Restoration, the Wetland Training Institute, California Society for Ecological Restoration, the California Department of Transportation, Parks Canada, and others, we would not have had the opportunity to refine the concepts presented in this book.

Don Falk, then executive director of SER, gave a helpful push for beginning this project, and James Aronson, editor of the Science and Practice of Ecological Restoration book series, showed tremendous patience as we sorted out our lives so this book could be finished. Equally patient were Barbara Dean and Erin Johnson of Island Press as they guided us through the publishing world. Thanks also go to John Anderson, Liz Cieslak, and Emily Allen of Hedgerow Farms for their suggestions on plant material. We also thank Carol Janis, Gladys Baird, Kent Askew, Tom Griggs, and Dave Strickland, for their comments and suggestions on the various drafts of this book, and Bob Allen, Haigler "Dusty" Pate, Keith Bowers, Ken Burton, and Michael Toohill, for their contributions in selecting and explaining projects to highlight.

Special thanks are due to Julie St. John and Elsa Hanly for their extensive comments and critique of the manuscript. Our gratitude is also extended to Andre F. Clewell for reviewing the manuscript and writing the foreword. We offer our sincere thanks to Mary F. Platter-Rieger for her photographic skills, technical computer support, critical assistance in preparing the graphics and photographs, preparation of the "References Cited" section, and editorial comments through the long journey toward completion. Encouragement and support from family and friends always present gave us the final energy to see this project completed.

Nothing is more satisfying to us than seeing a once derelict, degraded parcel of land return to a viable, healthy ecosystem supporting a diversity of life that was previously absent. What's even more satisfying is knowing that people from all different professions and interests around the world are restoring ecosystems and that we all work together as silent partners in a global movement to repair the planetary environment on which we all depend for sustenance.

As thrilling and meaningful as your participation in the restoration movement may seem, ecological restoration presents challenges and diverse options that could impede your ability to attain your goals. Your ingenuity and perseverance are required to restore impaired ecosystems. You have to use sound professional judgment to decide what interventions to apply and how to schedule your work in order to stay within budget.

Frequently, your concerns are compounded by impediments and constraints. Will your nursery stock arrive on time and in good planting condition? Are there sources of continuing environmental degradation on the adjoining property or upstream in the watershed that will negatively impact your recently completed project work and your reputation? Will residents in the local area respect your project or hinder your efforts? And then there are all of those logistical impediments. How do you prepare in advance for contingencies like extreme weather events and times when a labor crew never arrives? You may experience equipment failures. Wild geese may discover your newly planted nursery stock and perceive it as "lunch."

We give you a bird's-eye overview of what restoration projects are all about. We wrote this book as if we were in the field encouraging your efforts and sharing our experience. Clients and project sponsors need to know the level of effort required. Those who finance projects need to be realistic about the allocation of funds. Regulatory personnel need to know how to encourage effective restoration. Growers in native nurseries need to know why it's important to prepare stock for the particular environment of a project site.

This is not a cookbook with foolproof directions—no such book could be written that addresses the myriad issues you will face at a project site. Instead, this is a problem-solving book. We suggest a framework, process, strategies, and tactics that we have found to be effective. We lead you through each of the steps in restoration projects. We point out concerns and pitfalls that are common to many projects, and we take you through the thought processes that are germane to any project.

The term *ecological restoration* is defined by the Society for Ecological Restoration (SER) as "assisting the recovery of an ecosystem that has been degraded, damaged, or destroyed." We strive to perform ecological restoration in a thorough and lasting manner. Sometimes, for reasons beyond our control, the scope of our work may be too limited to perform ecological restoration according to SER's definition. Instead of nurturing wounded nature back to health, we sometimes have to apply rapid "emergency room" tactics and hope they lead to an eventual cure. Clients may promote projects as "restoration" when they would be more realistically designated as "rehabilitation," "ecosystem management," or "ecological engineering." In this book, we are not going to be sticklers on whether or not a project meets the strict definition of ecological restoration. Our main concern is to do as much as we can toward the eventual restoration of an ecosystem, even if circumstances prevent us from attaining a satisfactorily restored state in the short term.

The pace of restoration projects can be fast or slow, depending on the methods we employ. Most clients are eager for results and want to see a return on their investment. This is understandable, but it prevents us from considering more leisurely and less invasive strategies by which we minimally assist natural processes in the recovery of an impaired ecosystem. Such strategies have been playfully called "restoration lite" by restoration advocate Tein McDonald. It's the way that she and her compatriots in Australia commonly practice restoration on degraded public lands. North American clients are generally less patient. The projects and examples we describe in this book emphasize more aggressive tactics that satisfy our sponsor's and other stakeholders' concerns and those of regulatory authorities who oversee restoration projects.

In conducting restoration, we strive to reengage arrested ecological processes so that an impaired ecosystem can recover, adapt, and develop in whatever way contemporary environmental conditions allow. Attaining the processes, we believe, requires a results-oriented approach. We are not as interested in recovering a former state of an ecosystem as we are in ensuring a robust future state. Site-specific circumstances and environmental conditions may preclude the previous ecosystem condition from occurring, and you may have to clarify that distinction for clients and oversight personnel.

For these reasons, you should be well aware of what constitutes high-quality ecological restoration, as stated succinctly in the *SER Primer on Ecological Restoration* (available at http://www.ser.org) and at greater length in another book, *Ecological Restoration: Principles, Values, and Structure of an Emerging Profession* (second edition, Clewell and Aronson 2013), which is part of SER's book series on restoration with Island Press. These publications focus on principles of restoration more than on their application. They identify and explain the ecological attributes of a well-executed restoration project that returns an impaired ecosystem to health and allows it to resume its development as a continuation of its former ecological trajectory. As restoration professionals and as citizens of planet Earth, we should take advantage of opportunities to practice restoration of that quality whenever the opportunity arises. And as we work, we should strive to improve our profession in a collective global effort to repair our badly damaged planet.

Our principal contention in this guidebook is that an ecological restoration project should be a structured process and should proceed through an orderly pattern of operations. The four fundamental steps that we advocate in this methodological approach are (1) planning, (2) design,

(3) implementation, and (4) aftercare. These steps comprise a framework or road map that takes you from the initial conceptualization of a potential restoration project through the management of a completed project site. We provide checklists and flowcharts as aids along the way for project management. We emphasize project management throughout this book, because we know of many projects that have suffered because of ineffective management.

Conducting an ecological restoration project is a journey of discovery and problem solving. For that reason, we highlight how effective restoration requires a practitioner to play several roles. He or she must be a promoter of effective restoration strategies when discussing a new project with clients and oversight personnel. This requires frank discussions on funding, logistics, objectives, and potential outcomes. At the impaired project site, the practitioner changes hats and becomes an ecological detective, figuring out what the preimpairment ecosystem looked like, how it functioned, what caused impairment, and what would be the appropriate reference model on which to base project planning. Then the practitioner changes hats again to become an inventor of sorts, deciding how to recover the impaired ecosystem to its reference condition within budget.

Throughout the book, we emphasize the practice of ecological restoration. We do not address the field of "restoration ecology," defined as the theory and science on which the practice of ecological restoration is founded. Other books treat that subject thoroughly (see, for example, Falk, Palmer, and Zedler 2006 and van Andel and Aronson 2006).

In addition to serving the needs of the professional ecological restoration practitioner, the approach and tools presented in this guidebook will greatly aid volunteers, volunteer coordinators, and landowners who wish to restore degraded areas on private and nongovernmental organization (NGO) lands.

This guidebook is written for professionals who participate in a world of contractually bid and formally approved projects or NGOs receiving grants with accountability or in progress obligations. These are projects that are predicated on construction and installation documents and that are commonly permitted and subject to regulatory oversight. For that reason, we have reorganized relevant entries from the SER *Guidelines* and developed their content in much greater detail. In this procedure, we follow Clewell and Aronson (2013), who elaborated on the principles of ecological restoration that were stated succinctly in the *SER Primer* mentioned earlier. In large part, this guidebook represents a considerable expansion of chapter 9, on planning and evaluation, in Clewell and Aronson (2013), and in that manner these two books complement each other.

Project Planning

Part I provides the fundamentals of the project planning process. We start off with developing an ecological restoration project by describing the major phases of a project: project planning, design, implementation, and aftercare. Understanding the why, what, when, where, and how for each of these phases is critical to a successful project.

Chapter 2 focuses on project management to provide an overall understanding of how a project can be managed. It is a summary of project management practices in relation to an ecological restoration project. Many tools, checklists, tables, and series of questions have been provided to help you advance your project. Not all of what we provide will necessarily be of value to your specific project. A tool or method used for one project may not be suitable for another. Such is the nature of ecological restoration, in which no two projects are exactly alike. Similarly, the circumstances of a project typically differ, or during your investigations you may discover differences that were not initially apparent. You will have to make choices. Our aim is to provide you with sufficient background so that you feel you have made informed decisions.

One of the most challenging, and sometimes frustrating, aspects of a project is defining the project. Commonly, a project is "self-evident,"; however, many projects involve factors that require careful deliberation and that typically are discovered only after some investigative effort. Defining your project (chapter 3) requires understanding the project site, which will greatly enhance the vision you develop in formulating a project. For some large projects, your initial effort will be only the first of many projects in the same program. Budgets are a major governor of projects in terms of the money to pay for material or labor or the equipment to get the work done. This is where the process of developing a project becomes iterative, with adjustments on the "want" side to the "done by" or "paid with" side of the ledger. In the end, most projects reflect a compromise of some type, either from resources or from other identified constraints. You will discover constraints and opportunities as you proceed from one tool to another. This method will present you with potential options that can maximize your project's benefits.

Part 1 is the "decision governor" of conducting an ecological restoration project. It provides the guidance from which to proceed with design and implementation. How you define your project and establish the project requirements will be your reference when making subsequent decisions later in the framework process. Similarly, monitoring and evaluation are done using the foundation developed in part 1. You will understand your project more thoroughly by going through the whole process of developing it from beginning to end prior to actually implementing it on the ground.

Chapter 1

Framework for Ecological Restoration

The size range and complexity of ecological restoration are broad. The uniqueness of each project site makes it challenging to follow a recipe-book approach. We have found that following a four-phase framework provides a structured approach to an ecological restoration project that will greatly help you advance your project with a minimum of wasted time and resources.

Four-Phase Framework

The four sequential phases of project development are (1) planning, (2) design, (3) implementation, and (4) aftercare. This framework applies to restoration projects regardless of size, ecosystem, or location. The framework approach emphasizes that the restoration practitioner begins with the end target in mind. It is structured to provide a more disciplined approach to the planning and design process, whereby objectives drive the action steps of the entire development process. Each step of project development can be divided into a series of "aspects" of the project. Starting with project management techniques and advancing through to aftercare, the focus is always on results. The process of project development may involve a few to several steps, depending on the complexity of the project. Each step should be carefully evaluated and, when appropriate, should incorporate lessons learned from previous experience. Attention at this point in plan development will help to avoid repeating failures, and will increase focus on achieving results, before moving on to the next step.

We encourage using the lists, tables, and figures as a starting point for organizing thoughts, data, plans, and actions as the project progresses through the four phases of its development. Use the flowcharts, tables, and checklists to begin the process of developing a sound and thorough ecological restoration plan. The flowcharts will help you understand the relationships among the many steps in conducting an ecological restoration project. The checklists and tables will help determine the specific information required at each step in the planning process so that requirements and other project commitments can be properly identified in advance to permit the smooth implementation of the project. The Plan Review Checklist (appendix 8) in the implementation phase is the result of numerous projects over many years as well as input from several individuals.

Project Planning

The foundations of a project are established in the planning phase, which will greatly facilitate project development through completion. The process of coordinating with the stakeholders and obtaining a consensus among project sponsors is critical to the project development process. Building this foundation is crucial for the successful operation of the project. Whether it is practiced consciously or unconsciously, project management is the foundation of all successfully implemented projects.

We encourage developing a well-crafted mission, goals, and objectives for your project (chapter 3). Enumerating and clarifying the project goals has several advantages. They can be recorded and remembered for use in future projects. They can be communicated to team members, sponsors, stakeholders, key decision makers, and regulatory agencies and argued in these settings as appropriate.

Goals and objectives will form the basis of many decisions, starting with design strategies, design approach, plant materials, and installation schedule. The process of developing goals and objectives will require thoughtful evaluation and coordination among the stakeholders. Consensus among project sponsors is critical in the project development process. All permitting or other regulatory agencies should actively participate in this important initial step of the process. Agency-permitted projects often add additional constraints to a project with specific conditions that need to be considered as early as possible because they will directly affect the process of identifying and quantifying the evaluation criteria used to judge the performance of the restoration effort. In addition, the conditions may add features not previously considered and could impact the budget and materials required.

Initially, goals and objectives can be identified in a small group, with participants who are more knowledgeable about the project site or the circumstances. This is the time to explore the maximum opportunities, to test the "what-ifs." Brainstorming ideas will set the initial foundation for the considered discussion that will follow. To help the initial brainstorming, it is important to conduct an initial site analysis. Results of a site analysis provide the basis for action steps. Some actions will be short term, and the results will be immediately noticeable. However, a thorough approach to restoration also focuses on the actions that have long-term implications. Once the approach has been initially laid out for the stakeholders and project sponsors, it will now be possible to conduct a more thorough site analysis and a SWOT-C analysis (a process of evaluating various factors identified during site analysis; see chapter 4), which will allow the final goals and objectives to be developed. Once agreed upon, the goals and objectives are then finalized and will establish the project requirements from which project plans and actions can be created. This systematic framework approach encourages the restoration practitioner to examine all of the factors that are at work on the project site that can influence the outcome of the restoration effort. The site analysis process introduced in chapter 4 requires a thorough, analytical approach to understanding the forces at work on the project site.

Not all plans go as expected, and many designers are now applying the risk management process to restoration projects (chapter 2; appendix 3). A risk assessment outlines what could go wrong on your project. The resulting risk avoidance plan is the key element of the risk management process. This technique helps the project development team reduce the impact of issues that threaten the successful implementation of the project. The aim of risk management is to develop backup plans early in the project development process so that when something goes wrong, the team has actions that can be immediately implemented to minimize impacts to the project scope, schedule, and cost.

Project Design

Design, the second phase, is where one will encounter numerous choices. The path chosen will be determined largely by the decisions made during the project planning phase. There may be only a few design options to meet project needs on small sites with a single plant community; however, on larger projects with more than one plant community, both management and construction strategies may be needed. If the focus is on a target species—a species of specific interest to stakeholders—combinations of strategies to create the specific habitat elements for the target species may be needed. Various stages in the design phase will involve situations that require solid judgment—for example, the complexity and the variability of a site may not permit a precise distribution of plants by either sowing or planting. Chapter 6 provides general guidance principles and suggests ways to address specific situations. However, these situations usually require a good knowledge of the species behavior or ecology.

The development of goals and objectives will in turn guide your approach to the restoration. The four variations presented in chapter 5 will guide the decisions needed to develop a cohesive plan. The initial concept plan will address the management strategies, if any, and construction elements may be limited depending on the size or complexity of the project site. From concept to the actual design, an iterative process of examining the site with its goals and objectives will force modifications as one gets further into the design decisions and the specifics required to execute the project. We have commonly found that changes will still be necessary because stakeholders, especially regulatory agencies, will have additional observations. Negotiations and project management skills will be required to resolve issues and come to final consensus on the design. On more than one project, we have had to make some changes we felt were unwarranted, but we could not resolve the issues satisfactorily. Ideally, any changes will not affect the overall structure of your design. Because you will have been communicating with all your project sponsors and stakeholders, this disaster should not happen.

Following these modifications, the next phase of the process is to develop more formal project plans with specifications. If the project is primarily a management-oriented process, then having protocols developed and agreed to by your stakeholders will permit the operations to proceed. Invariably, there will be some construction elements associated with management-based projects. An example of an exception would be the use of prescribed fire to achieve a management objective. Any construction element should be clearly planned and described to ensure an efficient implementation phase.

Project Implementation

The implementation phase encompasses the full range of projects, including those lacking complexity to extremely complex projects requiring several additional elements typically found in major engineered construction projects. It will be up to you to decide which steps to include in the project plans and specifications to ensure the project is described in the way most likely to attain the anticipated results (chapter 9).

Depending on the project's size and complexity, you may require only some simple line drawings and tables to keep the project under control. You will have to decide on the level of detail required for your project. Management-based activities require a dramatically different set of parameters and instructions than do construction and installation activities. Written instructions are the basis for informing people on what, how, and when to conduct a specific activity; the "where" typically requires mapping or some other form of designation that has been established on-site, such as permanent markers.

On any one project there can be a wide range of construction and installation activities with an equally diverse means of execution. These conditions require not only mapping, diagrams, and plans but also specifications to ensure that the desired product can be executed with a minimum of damage to the remainder of the site. Clear instruction is crucial. However, even with numerous reviewers, misinterpretation is possible. An on-site, or at least on-call, person capable of answering questions or directing the work should be associated with the project. This person is a critical element in ensuring the results you desire. On-site inspection also ensures that products and the condition of materials are acceptable. Poor material or unacceptable plants and seed will seriously affect the performance of the project. Routinely, a maintenance period from ninety days to two years follows the construction and installation of plant material. The time period for plant establishment is concerned with survivorship of plants, irrigation (if required), and weeding to prevent competition during this vulnerable stage. Typically, the end of the plant establishment period is the completion of a project, which can now be closed out and passed along to the stewards or owners of the property for long-term management as part of a stewardship program.

Project Aftercare

In addition to routine maintenance activities and depending on the regulatory or sponsor directions, stewardship programs may be needed for projects beyond the time period it takes to meet the stakeholders' success criteria. In some instances, this may be because the plants have not reached self-sufficiency in the short time following installation. An ecosystem comprises many different elements, with some requiring many years, even decades, to fully develop. Some ecological restoration projects may be subjected to stressors that require attentive care until the ecosystem becomes strong enough to resist stressors.

The purpose of aftercare is to continue the restoration efforts begun during the project implementation phase and to help attain the desired trajectory of maturation. The type of ecosystem involved will determine the activity or management. Projects implemented under an agency permit or other regulatory requirement often require monitoring for three to five years, whereas some large or otherwise significant projects can have monitoring requirements that last more than ten years.

The maintenance and monitoring of a site is often a forgotten aspect of land management, and if not forgotten, it is frequently poorly resourced with insufficient staff hours and insufficient funds. External influences that go unabated can quickly deteriorate the condition of the restored area. Most projects require some form of nurturing activities following implementation to facilitate the development of the site and to ensure that stressors are minimized during the early stages. So, whether your aftercare activities include regular weed abatement or maintenance of barriers, such as fences or dikes, some form of postimplementation follow-up treatment is typically necessary to ensure that the reintroduced processes can continue to advance instead of being left in a state of neglect and decline.

Monitoring will document the progress of the site. Regardless of whether or not your project has permit or sponsor requirements, an important aspect of aftercare is site monitoring and documentation. We encourage

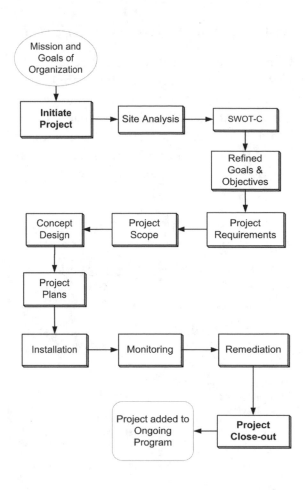

FIGURE 1-1. The project development process involves numerous tasks that each have inputs and critical products for a successful project.

all restoration practitioners to collect data and write progress reports to capture what is and is not working on the site. Our field is young, and the body of knowledge grows each year. We all can learn from even the simplest, most straightforward project. Communication among practitioners strongly helps to advance restoration practice.

We have introduced you to the four phases of an ecological restoration project and have discussed briefly the major actions taking place in each one (fig. 1-1). The chapters that follow will provide more extensive information and tools to assist in your journey through the restoration project development process.

Restoration Strategies

Ultimately, ecological restoration relies on the autogenic (self-sustaining) capabilities of the ecosystem. The spectrum of ecological restoration activities can logically be divided into two major strategies: (1) management and (2) construction and installation (fig. 1-2). Management-based actions are intended to reinitiate processes that would have occurred without the stressors on-site. This return initiates a host of autogenic responses. Construction begins with obtaining the necessary elements for doing the work and documenting what has been done. Following installation, a period of maintenance and monitoring ensures the persistence of the developing ecosystem.

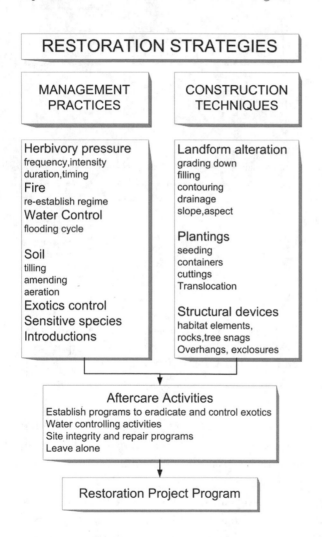

Construction involves the active fashioning of items put together by arranging or connecting an array of parts. Examples of construction and installation strategies include planting plants, installing rock features, and changing the grade or elevation of the ground. The specific strategy or strategies you choose will depend on the number of degrading stress factors affecting the project site, the time horizon of your project, your budget, and your project goals and objectives. Most projects require the restoration team to use a combination of both construction and management strategies to meet the project goals and objectives.

Management Strategies

Management strategies generally involve changes in present and past management practices and the use of a variety of management techniques. Management strategies typically include long-term restoration activities that don't necessarily result in immediately visible changes to the landscape. Land management varies with the local culture and resources involved. In the United States, fire suppression has been carried out to extremes, changing the structure and composition of vegetation communities dramatically. An interesting contrast to the suppression of fire in the United States is the routine fire

FIGURE 1-2. Restoration projects are the result of varying contributions of two strategies: management and construction.

FIGURE 1-3. The forest on the left illustrates the results of a prescribed fire regime, with a higher species diversity and a lower tree density. Cleveland, Ohio. (Photo by Lucy Chamberlain.)

practice and harvesting of heath for hundreds of years in Great Britain. The resulting ecosystem is composed of numerous sensitive and localized species. Here, fire is needed to keep the pines from taking over, an effort to maintain a specified succession stage. This strategy requires the site being degraded partially but not to the point where the natural resilience of the ecosystem is damaged beyond self-regeneration. By returning fire as an element within the ecosystem, nutrient cycling and other processes are promoted. Fire has been gradually introduced into many of the forests of North America to encourage regeneration of woodland species (fig. 1-3).

Another example of a management strategy is changing livestock grazing practices in the western regions of North America. Reduction of cattle stocking levels and exclusion or seasonal control of cattle grazing in riparian and meadow areas have allowed the regeneration of the diverse plant communities and native wildlife populations. Shifting the livestock controls overgrazing and secondary impacts, such as soil erosion alongside mountain streams.

Other management strategies include controlling water elevations or managing the rise and fall of water to enhance or impede species, and periodic weeding to reduce competition or to permit germination of native seed banks.

Sometimes restoration projects are undertaken for the sole purpose of addressing one species or element of an ecosystem. The practice of live trapping brown-headed cowbirds (fig. 1-4) throughout midwestern and western North America has resulted in population stabilization or increases of the Kirtland's warbler in the Midwest, of the black-capped vireo in Texas, and of the least Bell's

FIGURE 1-4. A modified Australian crow live trap on the Sweetwater River, intended to remove the nest parasite brown-headed cowbird. No nest parasitism on the least Bell's vireo for three years contributed to a 300 percent increase in population. San Diego, California. (Photo by John Rieger.)

vireo in California. In New Zealand, the practice of removing exotic predators from islands prior to reestablishing endangered or sensitive species has become a routine practice.

Construction and Installation Strategies

Construction and installation strategies for ecosystem restoration tend to be more resource intensive than do management activities. Construction activities typically include landform changes, such as slope and elevation, and the erection of barriers and temporary irrigation systems. Construction activities may include the removal of structures (e.g., dams or diversions) and the removal or rerouting of infrastructure (e.g., pipelines or roadways). A good example is the reconstruction of a stream that was previously "moved" to make land more usable (often for agriculture), which involves both the excavation of the "new" stream channel and the filling in of the rerouted stream channel. Construction strategies commonly require large sums of money to fund the activities and materials needed to restore the site. Whether you choose an approach that is more management oriented versus one that is construction oriented depends on the site and your established goals and objectives.

A study of a particular site might indicate that an element is missing that is largely responsible for the sustenance or continued existence of the ecosystem. Typically, this element is abiotic in nature, such as the alteration of hydrology. Levees around the bays and shores that eliminate tides are the most obvious situation. Land elevations may be lower than in the past, because of land subsidence or loss of soil caused by agricultural practices, or higher, as a result of sedimentation or fill.

Chapter 2

Restoration Project Management

This chapter will help you succeed in delivering restoration projects that satisfy the needs and expectations of the people and organizations interested in your projects.

Project management is the application of knowledge, skills, tools, and techniques to project activities to meet stakeholder goals and objectives (PMBOK 2008). Project management exists so that the project team can get the job done—on time, within the allotted budget, and according to the project sponsor's goals (scope) and objectives.

Project management begins with the end in mind and is concerned with achieving results that will satisfy your project stakeholders. Project management provides restoration teams or crews with a structured process that helps them coordinate their efforts so they create the right result, at the right time, within the targets established.

Project management helps you specifically define appropriate solutions to ecological restoration problems and helps you through the process of implementing them. Project management processes enable the project manager to integrate goals, anticipated site improvements, and environmental requirements into a set of project requirements. These become the detailed road map that the restoration team will follow throughout the entire project development process.

First, we'll begin by describing the four-phase life cycle of project development. Second, we'll introduce you to the five essential project management concepts. Third, we'll take a look at the elements of project management and examine how to build realistic project schedules and budgets. Fourth, we'll explore the concept of including a risk management element in your restoration plan. And finally, we'll close with some thoughts on how you can improve your effectiveness as a project manager.

The Four-Phase Restoration Project Development Life Cycle

Restoration projects share a common project development life cycle process that includes planning, design, implementation, and aftercare. The development process is product oriented: define, design, develop, and deliver (fig. 2-1).

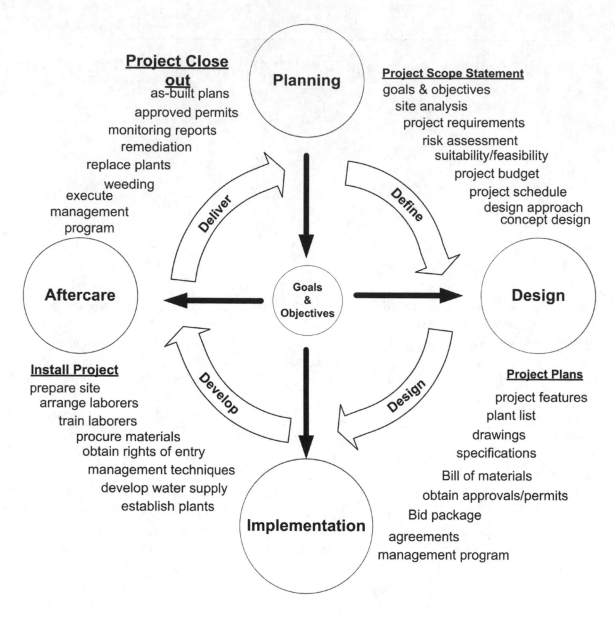

FIGURE 2-1. The four phases of project delivery include numerous activities and products. At the core of the cycle are the project goals and objectives. It is a circle to reflect the continued use of knowledge on subsequent projects or for rehabilitation of the same project.

This is in contrast to project management processes that are concerned with describing and organizing the work of the project. Each life cycle phase has a distinct process associated with that phase and results in a specific product (also referred to as a deliverable). A deliverable on a restoration project can be a plan (e.g., a planting plan) or an object (e.g., two hundred meters of pole barrier fencing installed) or even a service (e.g., surveys of warbler populations). After each phase of work is completed, the deliverables generated are used in the following work phase. For example, the project scope statement produced in concert with the project sponsors and other stakeholders during the planning phase forms the basis of design development work in the design phase. Another example is when project plans completed in the design phase go to a team for installation or execution during the implementation phase. Completion of the implementation phase leads to the aftercare phase. Products of the aftercare phase continue on beyond the project closeout. Lessons learned from this total experience of the project can be brought forward to your next project, thus connecting the life cycle back to planning, only for a different project. This model of a project can help explain the journey of a project to your stakeholders.

Each phase of a project has a distinct beginning and ending. It is the responsibility of the project manager to ensure that each phase of the project is completed as planned, with the expected deliverables.

The project management aspects of conducting a project include the following tasks: defining the project through goals and objectives in the planning phase; deciding on the strategy and details of the project, from quantities of items to protocols for management practices, in the design phase; acquiring all needed supplies, equipment, and biological material and performing the plans in the implementation phase; and last, meeting all objectives, totaling costs, and finalizing what got done, which completes the aftercare phase and so completes the project.

The Elements of Project Management

Project management entails three primary elements: people, processes, and tools. Project managers interact with these three elements to ensure successful project delivery. The people associated with the project are usually organized into project teams and are the ones who perform the work. The project manager leads the team, working with technical experts and laypersons alike in formulating and executing the project while soliciting and communicating with other stakeholders.

People: Project Manager

The role of the project manager is to anticipate and respond to issues before they become problems. The project manager must also respond effectively to internal and external stakeholders. Communication among the team and stakeholders is a critical responsibility. The project manager coordinates all of the various project activities to ensure they are completed as planned. He or she also ensures that all members of the team are trained to perform their tasks.

People: Project Stakeholders

A project stakeholder is anyone who has a vested interest in the project and its outcomes. Stakeholders on most restoration projects are grouped into one of three categories: external, internal, and other (PMBOK 2008).

The project team is composed of internal (volunteers, consultants, and contractors), external (individuals or organizations that will benefit from the project, such as sponsors and adjacent landowners), and other stakeholders (organizations that can facilitate or hinder the project delivery, such as special interest groups or governmental regulatory agencies representing the public interest in environmental protection). Naming or grouping stakeholders is primarily an aid to help identify which individuals and organizations can aid or hinder your satisfactory completion of a project. Implementing procedures for stakeholder input early during the project planning process can promote stakeholder "buy-in" to project goals and objectives and reduce or eliminate many problems that might otherwise occur later because of confusion over the reasoning behind decisions affecting project plans and construction.

External stakeholders are the individuals or organizations that will benefit from the improvements created by the project or will in some way be directly affected by the project. This group includes the person(s) or organization(s) sponsoring the project; the public, most often represented by individuals or organizations; nearby community members and adjacent landowners; and organizations or individuals providing finances or advocating the project.

Internal stakeholders are individual volunteers or organizations paid by the project sponsor that will use the deliverables, information, or enhancements produced. Such stakeholders include the project sponsor and the project team, the latter including consultants, contractors, and volunteers.

Other stakeholders are individuals or organizations that can facilitate or hinder the delivery of a project. Included in this group are affected special interest groups, regulatory agencies, and other governmental agencies, including governmental agencies that represent the public interest in environmental protection and natural resource conservation.

People: Project Team

The project manager works with, and through, the project team members to ensure that they deliver the project that is expected by the project sponsor and that meets the project goals and objectives.

For smaller, simpler projects, the project manager is also one of the major workers who execute a significant amount of the project products; however, in some cases, specialists may be brought into the process as needed. Most large and complex projects have multidisciplinary restoration project teams, typically composed of the project manager, functional specialists (e.g., biologists, landscape architects, ecologists, engineers, hydrologists, botanists, geologists, and soil scientists), and various consultants, contractors, and volunteers who perform project-related activities. Some project teams also include key stakeholder representatives. This is especially true when the stakeholders wield significant decision-making influence.

Depending on the complexity of the project, a four-member core project team—consisting of a project manager, a biologist/ecologist, a landscape architect, and a civil engineer—may be organized and tasked with completing all of the planning and design activities. Larger or more complex projects, or projects involving higher risk activities, usually call for a larger, more diverse team composition and a more closely coordinated project management effort.

Processes

Project management processes specify the deliverables required for the project, and they identify who will perform the work and when. Project managers rely on five project management processes to deliver the work items of the project. These processes—initiating, planning, executing, controlling, and closing—help the project manager form the project road map and keep the project on track toward successful completion (fig. 2-2). These processes are linked by the results they produce—the outcome of one becomes input for another.

INITIATING

The project manager ensures that each activity for the project begins as scheduled with the needed information and material required to get it done as planned.

PLANNING

The project manager plans the sequence of activities (work flow) for the project, including the effort required and who will do the work.

Typical planning tools include project goals and objectives, a scope statement, and a project cost estimate. Site analysis and risk assessment are part of the planning process.

EXECUTING

Concerned with delivering results, the executing process requires a coordinated effort among the entire project team. The project manager ensures that team members execute their work and helps the team maintain focus.

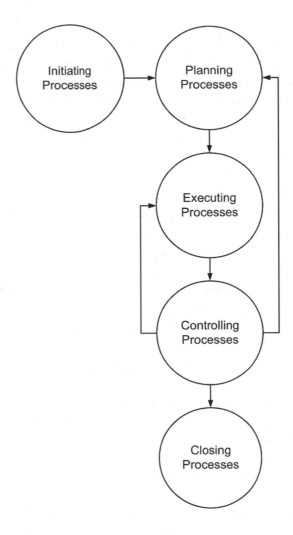

FIGURE 2-2. Successful ecological restoration project management makes use of five processes.

CONTROLLING

The controlling process is concerned with measuring project performance against the project requirements and taking any corrective action needed to ensure project delivery. Project control is proactive in its approach and also includes preventive action for initial phases as well as corrective action for the implementation phase in anticipation of possible problems (PMBOK 2008). Frequently considered an element of adaptive management, it is the way one can respond effectively to a discovered problem. Verifying that the team is doing the correct project work involves frequent monitoring and comparison with established project objectives. Effective project managers hold regular team meetings to monitor the status of key deliverable items.

CLOSING

The closing process, also known as closeout, is the final project management process. The closeout process brings the project to an orderly, formal conclusion. All project-related work is complete. Once a project is closed, no other project work is carried out.

Tools

The project manager's toolbox includes methods for developing schedules and budgets and for understanding the potential risks of a project. These methods are used by the project manager, team members, and others to successfully manage the project to completion.

Building Realistic Project Schedules

Project managers develop schedules so that they can plan and control all of the activities that make up the project. Project teams need project schedules to plan and execute the work to be completed on time.

Many types and formats of schedules are available. Martin and Tate (1997) identify four types of schedules: the milestone schedule, the deliverables schedule, the activity schedule, and the project schedule. The milestone schedule allows the team to take the goal of the project, divide it into subgoals, and assign deadlines to each. The deliverables schedule is used to show the delivery dates for all project deliverables. The activity schedule shows the duration and completion date for each activity required to create each deliverable. The project schedule has been most effective for our projects because it can show the entire project at a glance, and it is simple to develop and maintain. We also like this format because it is useful to internal and external team members who want to see the "big picture."

The project schedule is the most common schedule type used by project managers. It builds on the first three types of schedules described and includes start and end dates, the duration of each activity, and major milestones. Project schedules are commonly displayed in the Gantt format (sometimes referred to as a bar chart).

CREATING THE PROJECT SCHEDULE

To create the project schedule, the project manager should begin by listing each deliverable that has been identified in the project requirements as well as the person or resource responsible for completing that deliverable. Martin and Tate (1997) suggest using Post-it® Notes to identify each deliverable on the schedule and placing them onto a large poster-size sheet of paper (fig. 2-3). (There are several computer scheduling programs that can be used as well.) Once you've placed all the deliverables in logical sequence, you can connect related deliverables to create a network (flow) diagram. Deliverables that are required to be completed prior to the start of another deliverable should have a line with an arrow connecting them, showing the "before" and "after" relationship (fig. 2-4).

FIGURE 2-3. Developing a project schedule begins with identifying the various tasks and actions needed.

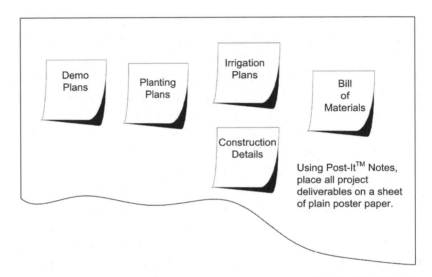

FIGURE 2-4. Arrange the tasks in a logical sequence; some tasks feed into more than one task.

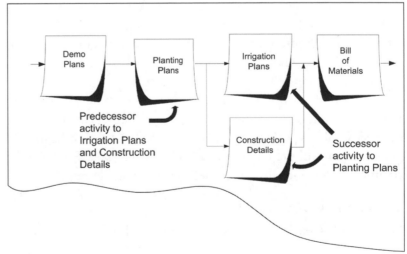

FIGURE 2-5. Following the
logic sequence, apply duration to
complete the tasks and review. The
longest path will identify the time
required to complete the project.

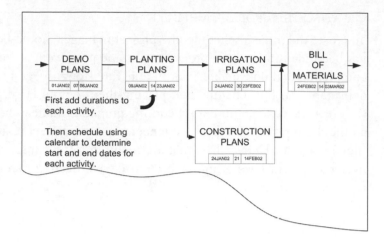

After all of the activity relationships have been defined, the time (duration) for completing each activity is noted on the Post-It Note (fig. 2-5); then, beginning on the left with the "start project" activity and moving to the right sequentially through each activity, the start and end dates are noted. Any critical dates known to the project manager, should then be integrated into the project schedule. For example, if there is a no-work window, such as a seasonal period when construction equipment cannot operate on the project site due to prohibitions on grading during the rainy season or concerns over impacts to wildlife (e.g., noise that might disturb bird species), then these restrictions should be built into the schedule.

Estimating the duration of project activities can be challenging, especially if you are estimating durations for activities that fall outside of your area of expertise. Fortunately, with ecological restoration work, much of the work required (e.g., planting, irrigation, earthwork, erosion control, and seeding) involves site work activities that are approached similarly to ornamental landscaping projects. For example, assuming that the specifications are the same, then the time it takes to plant five hundred 1-gallon-sized plants for a restoration project should be about identical to planting five hundred 1-gallon-sized plants in a park. Of course, site conditions, specifications, and other factors, such as the type of worker performing the work (i.e., volunteer versus paid laborers), will affect the actual production rates. However, if you look closely, you should be able to find a model project with enough similar conditions to enable you to use that work as a benchmark for comparing your estimate.

An alternative to determining activity durations is using the PERT (Project Evaluation and Review Technique; appendix 4). Interviewing others to obtain a series of estimates will provide the necessary data to perform this simple exercise. One significant effect of this technique is to reduce or at least control the tendency to "pad" the numbers. Such padding of the time estimates for tasks significantly extends the schedule, an unacceptable result.

THE GANTT CHART

Gantt charts clearly display the start and finish dates for each project activity. Activities (or tasks) can be arranged on the Gantt chart in any number of ways, including from highest to lowest priority or in order of finish dates or even by the person or group doing the work (box 2-1). The charts are usually used to provide a quick visual reference of the overall project status by showing the start and end dates for each activity. (The step-by-step process of making a Gantt chart is explained in appendix 1.) The power of a Gantt chart is that it is easy to create, easy to update, and very useful for both planning and control functions through the project life cycle. They are particularly useful when explaining the status of a project to a sponsor or other stakeholders. For most projects, a Gantt chart can adequately serve the purposes, but it may not be suitable for more complex projects or ones that involve multiple factors simultaneously. One weakness of Gantt charts is that they don't necessarily show the projectwide consequences of changing the schedule of a specific task—that is, Gantt charts typically only look at tasks as if they were independent activities and do not take into account their interconnected nature (Frame 1995).

Box 2-1. Gantt Chart Elements

1. List the activities required (work breakdown structure).
2. Estimate the time required for each activity.
3. Establish time units.
4. Identify the resources required.
5. Draw lines between the start and end dates.

Creating the Project Budget

The material expenses associated with constructing a restoration project can be daunting, especially if the site is in need of major improvements to sustain the planned restoration effort. Accurately estimating the overall project cost is important for project managers so that they can request sufficient resources to complete the project as scoped, and according to the project schedule. We have observed more than a few projects where resources apparently ran out prior to completion of all project activities, resulting in incomplete projects, stymied teams, disappointed project sponsors, and frustrated permitting agencies. The primary goal of project management is to deliver the project in accordance with the established scope, schedule, and cost for the project. Establishing an appropriate cost target at the beginning of the project is crucial for future project success. To develop an accurate project budget, it is important to examine two key elements: the project support budget and the project improvement budget. Taken together, these two elements form the overall project budget.

PROJECT SUPPORT BUDGET

The project support budget tells the project sponsor how much the project development component will cost and whether the project can be completed within the budget defined by the project sponsor. It is an estimate of the costs for all labor and materials required by the project team to complete all activities on the project, not including materials used in the implementation phase for site improvements.

The basis for the project cost estimate is the project deliverables and the activities required to deliver them. Using the Gantt chart, the project manager assigns the work to the responsible project team member. The project manager then estimates the resources (labor hours, days, weeks, or months as appropriate) required to complete each activity on the schedule. Using the unit cost per team member who will perform the activity (e.g., $85 per hour, or $13,600 per person month) determines the cost to complete the activity. If the activity requires more than one labor type (e.g., civil engineer at $150 per hour and biologist at $115 per hour), the hours and the costs are split accordingly. Once the total for each activity is determined, each activity on the project can be summed to determine the total labor component (appendix 2).

For any given project, certain material goods will be required to complete the activities. Any material costs associated with the delivery of an activity should be estimated and included in the budget. During the planning and design phases, material costs typically include such items as postage, photocopies, permits, plan reviews, mileage, and travel expenses. Typical office overhead expenses are not normally included in the budget as a separate item but are factored into the hourly rate charged for each labor type. All of the material costs required to complete each task are totaled, and then all of the material costs are summed to reach a final total for material goods.

Other items—such as a contingency factor, profit, and project management costs—are usually not included at this stage. These items, if required, are typically applied to the total project budget, including the project improvement budget (discussed below).

At this time, it is important to calculate the estimated cost of labor for establishment period maintenance, site maintenance, management activities, and any potential remedial actions. Frequently omitted are the labor cost estimates for monitoring and reporting over the length of the anticipated monitoring period. Projects requiring governmental permits frequently have specific designated monitoring durations. It helps to seek guidance regarding the anticipated required monitoring period from the regulatory agencies that will ultimately issue the permits required for project approval. Project sponsors need to understand that they will be responsible for funding these project components (i.e., maintenance, management, remedial actions, monitoring, and reporting) after the project improvements have been installed up until such time when the regulatory agencies sign off on their permits.

DEVELOPING THE PROJECT IMPROVEMENT BUDGET

Determining the budget for actual project improvements is a fundamental aspect of project management. Project managers are commonly asked to estimate the project improvement costs long before any restoration planning work has been completed. Resist preparing these shotgun estimates when you have little information to go by. Once you provide a budget estimate, you'll have an uphill battle with your project sponsor to increase the budget in the event you underestimated the work required. In some rare cases, when the proposed restoration effort is very well defined, the risks are extremely low, and the type of work required is well understood, then ballpark estimates such as cost per acre (per plant community or habitat type) might be useful. However, we have found that these types of ballpark estimates rarely end up being in the ballpark, let alone the same city!

We have also found that a certain amount of planning work must be completed prior to developing a reasonably accurate project improvement budget (box 2-2). Because of varying site conditions and project-specific features, ballpark estimating is just too risky. Certain specialty items (e.g., dewatering costs, vandal barriers, equipment costs, and remedial weeding) are difficult to estimate without some details about such issues as site conditions, the size of the area for the work, the season the work is to be performed, permit conditions, and so forth. There are too many uncertainties involved in restoration work to enable most of us to develop improvement budgets without first knowing some details about our project. To do otherwise is to invite yet more risk into an already uncertain practice. A particular axiom applies: the greater the guesswork, the greater the likelihood that the budget will change. Uncertainty in improvement budgets can be reduced, but only with good planning.

Box 2-2. The Bottoms-Up Approach to Budgeting

Developing realistic project budgets is best handled using a bottoms-up approach. This approach requires that you have some familiarity with the type of work proposed and the nature of the issues that you'll face with your particular site—and the more details you have, the better. Begin by using the project cost estimate worksheet shown in appendix 2.

1. Develop a model design that can be used to generate a preliminary bill of materials.
2. Identify the major project features that are to be installed on the project.
3. Generate an estimate of the quantities of each item on the bill of materials using the model design.
4. Apply the model over the proposed restoration area, and generate an estimate of the quantities needed for the entire restoration project.
5. Add in any specialty items that are not contained in the model (e.g., develop water supply, dewater streambed, repair stream bank, and so forth).
6. Total the quantities for each item of work.
7. Assign a unit cost to each item on the bill of materials (here, landscape architects and landscape contractors can be a big help in providing you with relevant data).
8. Estimate the labor costs required to build or install each item (landscape contractors may also be a good source here).
9. Include a list of all equipment needed to perform the work, the amount of time that you need each piece of equipment, and the equipment costs.
10. Total the costs for materials, labor, and equipment.
11. Add in a contingency factor to account for missed items and for undefined work. If this is the first pass of the budget, adding a factor of 25 to 50 percent of the total may be appropriate.

Risk Management in Restoration Projects

Risks to project completion abound in the field of ecological restoration. The potential for problems to occur in the restoration project development process is very high. However, certain strategies can be implemented to improve the likelihood of project success. Risk management is a project management tool concerned with avoiding or reducing risk as much as possible before it occurs. Good project managers create backup plans so that when bad things happen, the project team is not caught flat-footed.

Have you ever had one of the following problems?

- Planting is under way and an adjacent landowner informs you that you are blocking access to his land, causing the installation process to come to a halt.
- A key member of the project team becomes ill and is unable to complete design work critical to complete the project.
- Well into the design process, with installation about to begin, you learn that several species you required are unavailable due to poor seed set this year.

If you want to avoid these problems, you must get the people with answers to the above issues on your side from the very beginning of your project, and you must get them involved in an overall project risk assessment. By participating, these people contribute where they can be most productive: by identifying risks that the project will face and then identifying ways to counteract those risks. They are forced to come up with solutions for the problems that could potentially affect your project, and in the process they will likely become your most ardent project supporters, as they soon develop a strong sense of project ownership.

Restorationists can add value to their projects and their project sponsors by conducting risk assessments on their projects. We suggest that each project manager undertake a risk assessment during the early planning phase for each project. To do anything less is to leave the results of all the planning and design to chance and to risk making costly and avoidable mistakes. Risk assessment is a critical component in the restoration planning process; used properly, it can greatly improve the likelihood of on-time delivery and effective cost control.

A risk assessment for a restoration project should focus on answering the question "What can go wrong?" for each phase of the project. Carl Pritchard (1997) suggests that risks can be categorized by type to help identify and manage them. We suggest that restoration project managers focus on two types of risk: technical risks and environmental risks. Technical risks, which are primarily performance related, typically are concerned with the project deliverables and those team members responsible for delivering them. Environmental risks are concerned primarily with site and climatic factors.

Through brainstorming, your team should seek to uncover risks in both categories, evaluate the impact of each risk, evaluate the probability of the risk occurring, review the risks, and develop appropriate countermeasures to avoid or minimize the effects.

Each project will have its own assemblage of risks. Brainstorming with stakeholders and the project team will generate a list specific to the project. Environmental and technical risks frequently encountered on restoration projects include the following:

Natural Hazards
- Flood
- Fire
- Freeze
- Lack of precipitation
- Erosion (wind, water)
- Sedimentation
- Herbivory/browsing
- Insect infestation
- Vandalism

Management/Technical
- Changes in project requirements
- Changes in staff
- Limitations of team member skills
- Reliability of project design
- Maintainability of design (e.g., not self-sustaining)
- Availability of materials
- Political will
- Government regulatory involvement
- Inadequate coordination among team members and stakeholders

Schedule
- Unrealistic scheduling
- Labor/volunteer shortages
- Shortage of materials
- Failure to receive supplies due to supplier problems
- Miscommunication to suppliers
- Productivity of labor force not at expected levels
- Unforeseen site conditions
- Start-up difficulties
- Poor weather affecting installation or volunteer attendance
- Agricultural pest control activities

Cost
- Schedule delay
- Lack of understanding
- Underestimated project funding
- Incorrect calculations
- Vandalism
- Inflation of material (purchasing less)

- Changes in project requirements
- Failure of donors to follow through with commitment

Note that some of the risks listed here affect more than one category. It is not important to have these precisely grouped; it is more important to examine all aspects of the project for risks. Our experience in risk management has led us to recommend the following risk management steps:

1. Identify all known risks that could likely occur.
2. Assign an impact rating to each factor, using a scale of high, moderate, and low to the project scope, schedule, and cost.
3. Assess the probability, or likelihood, that the risk factor will occur sometime during the project span.
4. Assign an overall risk rating (risk rating = impact × probability).
5. Rank the risks, ranging from high-impact/high-probability risks to low-impact/low-probability risks.
6. Develop countermeasures (only for the high-impact/high-probability risks) that will help to avoid or reduce risk. Depending on the number of risks, consider choosing only the top eight so that you can properly handle the issues.
7. Select the backup plans (i.e., "plan B") for implementation.
8. If, after going through this process, the team cannot reduce the risk to an adequate level, then bring the issue to the project sponsor.

An example of the previous process follows: A key native plant supplier will not deliver *Salix* spp. by March: risk impact, "high"; probability of occurrence, "high"; risk ranking, HH. Risk management countermeasures generated during a team brainstorming session produced the following ideas: seek alternative suppliers; collect cuttings and establish growing ground on-site; delay project start until supplier can deliver; defer *Salix* planting until August; revise plant list.

All of the information generated in developing a risk management plan can be put into a worksheet that organizes the information into an easily understandable format (appendix 3).

The project manager who leads their team through the risk assessment process and builds understanding of what potential problems might occur will usually gain consensus about how the team will work to avoid the problems, thus increasing the project's prospect of success. Of course, not every risk issue can be identified or prevented, but you will at least be prepared to deal with the pitfalls that are avoidable.

Improving Your Results as a Project Manager

Being a project manager is challenging. To succeed requires becoming adept at establishing project requirements and communication. The following four principles of good project management are paraphrased from J. Davidson Frame (1995), in his book *Managing Projects in Organizations: How to Make the Best Use of Time, Techniques, and People*:

1. Be a deliberate project manager; don't assume the project will adhere to your plan.
2. Projects are complex and unique goal-oriented undertakings requiring careful planning and execution.
3. Take the time to plan at the beginning. It will improve the project outcome and will reduce the potential for costly rework, which wastes resources and increases project costs.
4. Think ahead and anticipate that problems will arise. Have contingency plans prepared in advance so you can put them to work immediately to reduce delays.

Simply knowing the tools and going through the mechanics of using them is not all there is to being an effective project manager. It also requires that you identify your stakeholder's needs so that you can define the project requirements and build a project that achieves stakeholder objectives. An effective project manager embraces the following responsibilities:

- Advocating for the project
- Managing the project team
- Working with the team and stakeholders to define the project and determine the schedule and cost
- Tracking the project progress and managing needed changes
- Facilitating resolution of problems and other issues involving the scope, schedule, and cost of the project
- Presenting project progress and issues to all stakeholders
- Receiving and evaluating stakeholder comments
- Coordinating a lessons-learned evaluation with the team
- Evaluating status and performance of team members and making changes where appropriate

Establishing Project Requirements

A project's requirements are essentially the combination of (a) the stakeholders' vision of the completed project, (b) the goals and objectives for the project, and (c) the list of anticipated actions, including all environmental permit requirements. Translating these inputs into clear and specific project requirements is the first major milestone for the project development team. Project requirements are the "what" of the project, describing the actionable items to be delivered to the project sponsor. This is truly one of the most challenging aspects of project development, as compromises and substitutions typically are necessary. All members of the team should have access to the report.

When establishing project requirements, consider these guidelines:

1. Be results oriented.
2. State the purpose clearly and concisely.
3. Use action statements (e.g., using such verbs as *create, develop, monitor, establish, install,* and *remove*).
4. Include known milestones (deadlines), such as start dates, end dates, and mandatory finish dates.

5. Be realistic.
6. Include illustrations (e.g., drawings, tables, and charts) wherever possible.
7. Obtain stakeholder approval (have stakeholders sign off on the list of requirements).

IDENTIFYING STAKEHOLDER GOALS

All projects exist to satisfy certain stakeholder goals. When the stakeholders' goals are satisfied, the project can be judged a success (Martin and Tate 1997). Therefore, it is imperative that project managers focus the project development team on answering the following questions:

- What is the project supposed to accomplish?
- Has the stakeholder identified any goals or objectives for the project?
- Can we resolve the need with restoration techniques?
- Who are the stakeholders most directly affected by the need?
- Do the stakeholders agree that the need is worthwhile?
- Are there policies or restrictions that may limit or impede the restoration efforts?
- When is the project to be completed?
- How should it look?
- Do the stakeholders have the same expectations?
- How soon should results become apparent?
- Are there special considerations that may govern the implementation methods or project schedule?

Identifying answers to these questions is one thing. However, managing stakeholder expectations is difficult because project sponsors and other stakeholders can have very different perspectives that may conflict with one another. Finding appropriate resolutions to such differences is a major challenge of project management. The project team must identify the stakeholders on a project, determine their needs and expectations, and then manage and influence those expectations to ensure a successful project.

HANDLING CHANGES TO PROJECT REQUIREMENTS

Invariably, project requirements will change. Therefore, each project manager should establish a system to carefully monitor and control changes made to the requirements. As a part of handling any proposed change in requirements, the project manager should document the change and seek approval from the project sponsor. This is especially true if the change significantly affects the scope, schedule, or cost. Depending on the magnitude of the change, the documentation can vary in information. At a minimum, the change should describe the action—including its impact to the project in terms of cost, schedule, or scope—and have the sponsor or other appropriate stakeholder sign and date the document. If the change is significant, it would be helpful to the team to list the tasks and staff affected and provide a detailed schedule and cost analysis.

The change documentation should include the following information (Frame 1995):

- Date of the change request
- Name of the person requesting the change

- Description of the change
- Statement of the change's impact on the project
- Listing of tasks and staff affected by the change
- Estimate of the cost of the change
- Signature of the individual making the change request, indicating awareness of the change's impact to the scope, cost, or schedule
- Evidence of stakeholder approval of the change (have stakeholders sign off on the change request)

COMMUNICATING CLEARLY AND THOROUGHLY

Frequent, understandable, and concise communications are essential to move toward successful completion of the project. It can be very useful to create a communications plan for each project. This plan should include details of the information to be distributed (e.g., format, content, level of detail), what kind of information it is (e.g., status reports, data, schedule, technical documentation), what methods will be used (e.g., meetings, written reports), and a schedule for production and distribution of the information.

The most common forms of communications include meetings, telephone and conference calls, formal letters and memos, fax transmittals, e-mail, and Internet. E-mail is a very fast, inexpensive, and efficient way of distributing information. A website can also be used to send e-mail to organizations outside the project team, sponsor news articles on the project, disseminate information on the project, and provide links to local interest groups participating in the project.

The obvious purpose of team meetings is to communicate information. A less obvious function is to establish a concrete team identity and build relationships. During the meeting, team members get to see that they are not working alone; rather, they are part of a larger group, and the success of the group undertaking depends on the efforts of team members doing their individual parts. Typical types of project meetings include the following:

- Kickoff meeting—a meeting that sets the stage for the project team's roles and responsibilities
- Status meeting—a periodic meeting held to examine project performance and to discuss issues
- Lessons learned meeting—a meeting held usually during the closeout phase to discuss what went well and what needs improvement

The lessons learned meeting is commonly overlooked in this process. After the team has completed its work, getting the team members together to discuss the project can be challenging. However, if you or your organization plans to continue conducting restoration activities, it is strongly recommended that this meeting be conducted. Restoration work can be challenging enough; when mistakes are made or unplanned events happen, the team's response and feedback can be valuable information. If the unexpected occurrences and the responses to them are not documented, it is possible—whether because of changing personnel and simply the passage of time—that important actions can be forgotten.

Productive meetings of any type involve the following elements:

1. Create an agenda.
2. Record the action items.
3. Identify the person or persons responsible for each task with a due date.
4. Record all decisions.
5. Document items for future discussion (we refer to these as "parking lot items")—issues that come up during a meeting that are not relevant to the current agenda items but will be important to take up at another time (Hindle 1998).
6. Assign a facilitator. For typical meetings, this can be handled by the project manager or another designated person.

A facilitator can aid hugely in obtaining needed information and providing guidance for heavily attended meetings or ones for which complex issues are anticipated. Also, keep in mind that meeting minutes do not have to be elaborate or long. In fact, the more succinct the writing, the easier it will be to keep track of the tasks that have been assigned.

Chapter 3

Defining Your Project

A restoration project commonly begins with a great deal of enthusiasm, high expectations, and a strong sense of camaraderie among team members—an understandable response to the exciting idea of restoring the land. However, not long into the process, the team may discover that perhaps they don't all share the same vision of what needs to be done. A simple project of invasive weed eradication, for example, may largely avoid this potential conflict, unless the invasives have made significant inroads into the plant communities and leaving large open areas of bare land is unacceptable. You have now entered into the arena of potential conflict due to varying expectations among team members.

Fortunately, the project definition process helps create a common understanding among all participants in the restoration project. As mentioned in chapter 2, these participants are more than just the people working on the ground. Collectively, we call the larger group of interested parties and organizations "stakeholders." The process of reaching agreement on what is to be accomplished results from all participants expressing their individual visions of how to resolve the problem(s) (chapter 2). All of these points of view are coalesced into a mission statement representing the consensus of the stakeholders, which is then expanded into a set of project goals. Derived from these goals are objective statements, which in turn generate specific actions that will be implemented to achieve each project objective.

This step-by-step planning process ultimately results in clarity and consensus. The mission statement, goals, objectives, and actions serve to define a restoration project. An important function of having a clear project definition is to prevent straying from the agreed-upon scope of the project as set forth in the mission statement and established goals. This will help to avoid wasted efforts or loss of resources and to prevent misunderstandings if some stakeholders show up at the table late in the planning process.

Restoration project managers understand from the beginning that forces are at work that will challenge their stated definition of the project. During the project development process, you may face budget shortfalls, changing project requirements, shifting opinions of project sponsors, and unforeseen environmental and site conditions (see the list of risks in chapter 2). That is why it is so

33

important to develop a procedure for decision making that will help you manage the project scope so that you can ultimately deliver the project that is expected by the stakeholders.

Project Definition

Project definition fully describes the project to be performed in terms of what, when, and where. In addition, the description includes specific actions, the budget, and the overall schedule. The end result is a document that can be used as a reference by members of the team, other stakeholders, and the general public.

Beginning the project-defining process requires conducting first a preliminary site analysis (chapter 4) and then a visioning exercise known as leitbild (discussed below). Ideally, these initial data will validate the initial mission statement for the project and provide structure to the preliminary goals. Later on, further refinement of the goal statements will follow, with more detailed site analysis and use of reference sites (chapter 5). This iterative process is necessary because options commonly do not present themselves until additional information is discovered on-site or with further research.

Project definition begins with stating the preliminary project goals and objectives. The process of establishing goals and objectives usually results from inputs by the project sponsor and other key stakeholders and is a response to the preliminary site analysis of the project site. A reference model derived from data obtained at a reference site will guide the decisions on goals and objectives. However, situations do occur when no reasonable reference site is available and alternative approaches for developing a reference model will be needed (chapter 5). A clearer vision of the restored landscape begins to emerge as the project moves through the project development life cycle, during which more definition is provided by team members.

The project goals and objectives are tested and refined during the site analysis process when the project team evaluates site function and determines specific requirements that must be addressed by the project (box 3-1). This process may require initial field trials to verify suitability (box 3-2). Finally, a project scope statement summarizes the definition process and documents the overriding design requirements to which the project must adhere to be considered successful.

Function

Because of the complex sets of interactions that occur within an ecosystem, it can be difficult to isolate a single process for study. Yet, this is the responsibility and challenge facing the restorationist because the absence of a function (or functions) within an ecosystem can profoundly affect the overall health and long-term sustainability of that ecosystem.

Put simply, a function is the combination of all the natural processes and events that occur within a given ecosystem. These processes can be divided into biological and physical categories. Some common physical processes and events include day length, solar access, temperature, erosion, and water percolation. Biological processes include growth, mortality, decomposition, competition, predation, herbivory, parasitism, and symbiotic relationships. Events tend to be singular in nature, and they characteristically have distinct beginnings and endings; often, events trigger

Box 3-1. Restoration Project Highlight: Test Trials of Restoration Approaches Can Reveal the Most Cost-Effective Methods for Large Degraded Areas

Project: Invasive Tree Removal and Wetland Restoration in Barataria Preserve; Jean Lafitte National Historic Park and Barataria Preserve, Louisiana, United States

Before the current Barataria Preserve was established, the Jean Lafitte National Historic Park contained numerous construction activities—ranging from a failed residential community to oil exploration to pipelines—that altered the wetland topography and interrupted the wetland hydrology. A series of canals that had been excavated for pipelines, oil exploration, and various other activities occupied some twenty miles of waterways. The spoil from the excavations was commonly cast on either side or both sides, depending on the purpose. These spoil banks ranged in width from twenty to eighty feet with a height of from two feet to almost eight feet.

In effect, an upland habitat was created in a wetland as a long linear habitat crossing a large portion of the preserve. Invasive plants were able to establish in these new upland areas, with the Chinese tallow tree the most dominant. But the effects of the invasive species were not as significant as the interruption of the water flow through the wetland. The spoil banks created numerous locations where the altered hydrology seriously impacted the native vegetation and function of the wetlands.

Louisiana State University, working with the National Park Service, conducted test trials on the most effective way to remove the spoil banks and fill the canals. The trials involved testing two approaches. The first approach was to place spoil into the adjacent canal with supplemental material dredged from a nearby lake, in an effort to return the watercourse to its original elevation. The second approach was to use only the spoil material and leave the watercourse in a shallow but not original condition. Monitoring of the two channels provided data for future restoration work. Although some differences existed between the two approaches, the decision was made not to import any further dredge material. Factors leading to this decision included the added cost (approximately eight times more expensive) and various logistical issues of transport and access. When funding became available through the American Recovery Administration, the preserve was able to obtain sufficient funds to convert 4.4 miles of canals to shallow water habitat and reestablish normal wetland hydrology to that area of the preserve.

The outcome of the construction went as anticipated; however, not anticipated were impacts to various visitor groups in the preserve. The preserve is open to hunting, trapping, and fishing as well as recreational boating. The construction work did interfere significantly with hunting activities. The tree removal, spoil bank excavation, and partial refilling of the canals was deemed an overall success, and the preserve is seeking more funds to work on the remaining ten miles of canals.

Box 3-2. The Benefits of Goals and Objectives for Ecological Restoration Projects

- Defines what project success should be
- Defines what the project is all about and what the project is not about
- Forms the framework for the restoration project planning and design
- Helps to define the strategy for project implementation
- Helps to promote efficient use of resources by focusing the project development team on activities that will achieve the desired results
- Forms the basis for postproject evaluation

---

processes. For example, events such as floods, fire, and drought all tend to precipitate processes that affect both the physical and the biological levels.

Understanding and interpreting an ecosystem and detecting the missing or poorly performing functions are the most important tasks in conducting restoration work. This requires a working knowledge of the ecosystem types you are attempting to restore. The ability to discern the interdependencies and interactions of the various elements that make up a function comes from observing both healthy and unhealthy ecosystems.

Global concern for wetland loss and quality has generated numerous studies addressing functions of wetlands. Although some functions are unique to wetlands or wetland-type conditions (Hammer 1997), most of the functions are also present in upland situations (Tongway and Ludwig 2011; Tongway 2010; Friederici 2003). The following list introduces a range of common functions identified on restoration project sites.

- Transformation of nutrients (primary production)
- Element cycling, including carbon sequestration
- Removal and retention of nutrients and compounds
- Maintenance of local gene pool
- Maintenance of plant populations (diversity)
- Maintenance of animal populations (diversity)
- Maintenance of endangered species (plant and animal)
- Resilience (recovery from disturbance)
- Resistance to invasive species
- Resistance to herbivore outbreaks
- Pollination
- Support of food chains and webs
- Flood hazard reduction
- Erosion control
- Stream bank and shoreline stabilization
- Water retention
- Groundwater discharge
- Groundwater recharge

- Access to refuges during high water
- Provision of habitat for dependent species
- Maintenance of habitat connectivity and ability to disperse
- Land use buffer

As discussed in chapter 2, the first step in beginning a restoration project is to develop a problem definition. Your objective is to identify missing or underperforming functions. Using information collected from site visits and various documents enables the condition of the various attributes of the ecosystem to be considered. Next, those elements requiring improvement or reintroduction are noted and a list of potential solutions is developed.

A simple tabular worksheet (table 3-1) is an effective tool for organizing initial thoughts about a specific project site. This phase of project planning will be an iterative process as additional information is discovered.

Table 3-1. Site Inspection Worksheet

Existing Condition	Desired Restored Condition	Possible Actions Required
1. Site densely vegetated by several weed species	Weeds having a minor role in terms of biomass and cover	Remove weeds manually; use herbicide; provide plants that will create a denser cover to compete with weeds; cover ground with mulch; burn weeds prior to seed set; exhaust seed bank
2. Localized erosion rills and gullies	Stable surface slope capable of tolerating surface runoff without mass soil movement; increased relief to slow water movement and erosive energy; avoidance of concentrating drainage and buildup of energy	Stabilize surface by compaction and mulching
3. Lack of diversity of animal populations	A 50 percent increase in mammal, reptile, and bird species using data from the existing site; more diverse microhabitats	Introduce rocky piles and deadfall, dense vegetation patches, and possibly water feature, if appropriate to location
4.		
5.		

At this stage in the process, you will have the basic elements leading to a restoration plan. This general idea of the condition of your restoration site and what you want to restore will continue to develop, modify, and change as you go through the site analysis process (described in chapter 4). Only after you have gone through the site analysis process will you be able to fully refine your goals for the site. The actions you identified in your worksheet can be refined, and the associated objective statements can be more precisely crafted. The goal and objective statements will be the critical foundation for any restoration plan.

Leitbild Concept

Leitbild, a German word meaning "ideal" or "guiding principle," is a "frame of mind" for planning a restoration project. It is a means of visualizing and developing a project without considering any constraints on the project design (Middleton 1999). Leitbild originated in Germany (Larson 1995) for use in stream and river restoration and is currently being applied in numerous European countries. When embarking on a project, one often has a dream or visualization of how the site will look based on one's knowledge of the area and the expected changes that the various ecosystems typically undergo. Most likely this vision encompasses a condition that is devoid of human structures or modifications and is not limited by lack of resources.

In essence, leitbild is an exercise in response to the often asked question, What would you do if you had all the time needed, all the money required, and no outside restrictions? Using available data, your team can develop an "ideal," which will then undergo an iterative process of modification as physical, biological, fiscal, legal, and political factors, in addition to resource constraints, are identified and their influence accounted for in the restoration proposal (Valentin and Spangenberg 2000; Muhar, Schmutz, and Jungwirth 1995). This vision is constrained by the need to restore the attributes of naturally functioning ecosystems. The leitbild process results in a realistic project that has been tempered with the realities of the project site and resources available, while maintaining the historic continuity of the ecological trajectory of the site. The leitbild is not a means of engineering the ecosystem but can be involved in reverse engineering a site. Leitbild is a tool you can use to explore all options for your project.

Developing a leitbild can be an important first step in the process of developing goal statements. Using a leitbild approach enables you to consider the full range of possibilities for your site. Benefits of the leitbild method include the following:

- Focuses attention on what is best for the site
- Leads to more comprehensive planning and design decisions
- Helps the restoration team to better visualize the project goals
- Causes the restoration team to think not only at the project level (short term) but also at the program level (long term)
- Creative method of problem solving that focuses attention on what is possible and not on what is feasible

Explanation of Steps in the Project Definition

All projects begin with a need. In the case of ecological restoration projects, the need is to return ecological processes to a site. The process of developing a viable ecological restoration project begins with the problem statement that identifies the concerns. Following the problem statement, a mission statement is developed and begins the process of focusing on a project that will eliminate the perturbations on the site and restore missing elements. Next, data (chapter 4) are collected to understand the site and develop a series of goals that focus on achieving the stated mission of the project. The goals form the basis for developing objective statements, which are followed by actions. Objective statements are specific, and actions are very specific.

Project requirements can be formulated from the prior steps and become the basis of future project development actions. A scope statement can be developed to serve as a guide for the restoration team and inform others about the parameters of the restoration effort. A scope statement includes more than just the goals, objectives, and project requirements but also the project schedule and budget. Agreement on these elements by the various stakeholders is highly desirable. However, as discussed in chapter 2, project stakeholders have varying degrees of involvement. Depending on the project, site, and resources involved, some stakeholders are advisory and some have legal or fiscal involvement. Sponsors of the project, typically, the entity that is funding the work, will want to be confident that the reasons for their involvement are satisfied by the restoration work to be performed.

Problem Statements

All restoration projects should begin with identifying the concerns related to the ecological health of a specific site. This statement of concern for a site can be written initially by a concerned individual, committee, or organization. Frequently, citizens will form groups to initiate action for correcting the degradation of a site. This group may approach a governmental body to have some action taken. This initial group develops a statement of the problems and identifies deficiencies at the site to inform the decision makers so that they recognize the need for action. Problem statements are clarifications of concerns as to how they relate to a specific resource value or set of values (e.g., eroding stream banks or lack of riparian habitat for birds).

Here is a sample problem statement: "The current density of invasive vegetation in riparian vegetation and barriers along Kaskaskia Creek has resulted in low populations of riparian-dependent wildlife species, especially birds."

Mission Statements

A mission statement is a concise statement clarifying the special task a group is to perform or the purpose for which an organization has been created. A mission statement is visionary, relatively general, and brief. Some restoration projects are just one component of a larger program for achieving the overall mission of an organization or agency. It is possible that the missions of some organizations and agencies do not fully embrace ecological restoration. A conflict of desired outcomes may result among the different stakeholders. To avoid wasted resources or an outcome that fails to satisfy the stakeholders, this situation must be resolved before proceeding.

Here is a sample mission statement: "Restore the natural landscape and habitat conditions to how they were when the ranch was in operation."

Goals, Objectives, and Actions

Establishing goals and objectives is the most powerful and defining act in the restoration planning and design process: powerful because no other planning product sets in motion so many subsequent decisions and follow-on activities; defining because objectives, once invoked, become touchstones against which all subsequent project results are measured. Successful restoration project development begins with clearly stated goals and objectives. These statements are then translated into specific project requirements, which are actionable items that the project team uses as a road map to guide the project design (box 3-3).

Goals are expressed very early in the planning process and serve to define the overall purpose of the project (Doyle and Straus 1976). A goal is a complete statement of purpose or direction; it is general in nature, broad in scope, and flexible enough to persist over time. Goals describe expected results—they paint a picture or a vision of a desired state or condition. They are stated in clear, succinct, and easily understood terms and are as inclusive as possible, meeting the interests of a broad-based group of project stakeholders. Effective goals serve the following purposes:
- Enable us to focus on achieving desired long-term outcomes rather than starting out trying to resolve "urgent" problems and issues
- Promote consistency of perspective among the many different stakeholders involved in the planning process
- Enable further uniformity of requirements and actions among overlapping governmental jurisdictions
- Form a basis for testing the validity of suggested objectives, actions, and programs in the context of meeting the stakeholders' vision for the ultimate state or condition of the resources

All projects exist to satisfy certain stakeholder goals. If needs are not seen to exist, no action will be taken to satisfy them. Martin and Tate (1997) say that if the stakeholders' goals are satisfied, then the project can be judged a success. Therefore, one imperative for restoration project managers is to focus the project restoration team on answering the following questions:
- What is the project supposed to accomplish?
- Has the stakeholder identified any goals or objectives for the project?
- Who are the stakeholders that are most directly affected by the need?
- Do the stakeholders agree that it is a worthwhile need?
- Can we resolve the need with restoration techniques?
- Are there policies or restrictions that may limit or impede the restoration efforts?

Box 3-3. Restoration Project Highlight: Successful Restoration Projects Often Must Address Multiple Issues

Location: South Cape May Meadows Restoration Project, New Jersey, United States

Cape May, a small community in New Jersey, historically was subjected to flooding on a regular basis. The citizens made their case to the US Army Corps of Engineers, which was not able to justify doing work based on the traditional cost-benefit analysis. However, as the area received additional addition, it was decided that more than a flood prevention project was needed. In concert with the State of New Jersey and The Nature Conservancy, a project was developed that provided shoreline, dune, and wetland restoration within this complex.

The residents of Cape May were primarily concerned with keeping the ocean from flooding their community, which would require modifying the shoreline and repairing the sand dunes. The wetlands were also seriously impacted by the effects of an abandoned village with a grid of roads creating square ponds and wetland segments, which created another problem by allowing phragmites, an invasive species, to invade the marsh. The hydrology had also been altered, which negatively affected the adjacent landowner, Cape May Point State Park. The restoration project developed by the US Army Corps of Engineers and the New Jersey Department of Environmental Protection included several elements. The most significant of these were beach replenishment and dune reconstruction, the restoration of freshwater flow through the wetland by re-creating an old stream channel, and the construction of levees and installation of water control structures.

The Nature Conservancy and the State of New Jersey have developed a water management plan to allow water movement through the property while retaining appropriate water levels at various times of the year to enhance migratory bird habitat. No flooding has occurred since the work was completed in 2007, and the invasive species were eradicated to less than 5 percent of their original area. The Nature Conservancy will take over active management of the Cape May Meadows in 2013. Professional staff will be responsible for controlling invasive species and managing water levels. Although the primary concern of adjacent landowners was to control the flooding, they soon realized the positive benefits of this restoration work. With more than two hundred thousand annual visitors, this site is now a significant natural attraction.

- When is the project to be completed?
- How should it look?
- Do the stakeholders have the same expectations?
- How soon should results become apparent?
- Are there special considerations that may govern the implementation methods or project schedule?

Identifying answers to these questions is one thing. However, managing project sponsors and other stakeholders can be challenging because they commonly have very different perspectives with goals that may conflict with one another. Finding appropriate resolutions to such differences is a major challenge of project management. The project team must identify the stakeholders on a project, determine what their needs and expectations are, and then manage and influence those expectations to ensure a successful project. In general, differences among stakeholders should be resolved in favor of the project sponsor. Of course, one significant challenge is to ensure that whatever the concessions are, the project still conforms to the attributes of ecological restoration.

Whereas goals are strategic, objectives are more tactical in nature and are the means by which goals are brought into reality (fig. 3-1). Understanding the differences between and relationship of these two words will serve you well in the planning phase of a project. Understanding and using objectives is SMART (box 3-4).

Consider the examples of goals and their associated objective statements for the following sample project, Cowden Preserve. Keep in mind that goals are attained, not measured, and objectives are specific and measurable.

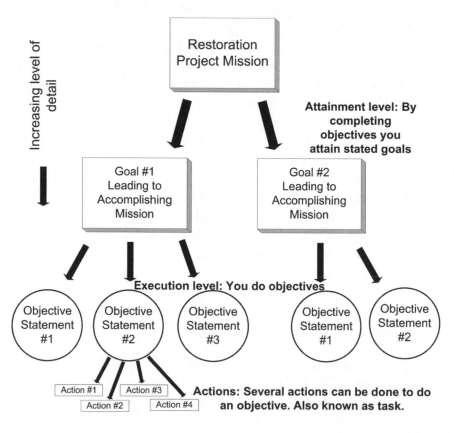

FIGURE 3-1. Understanding the relationships among action, objective, and goal is critical to a successful planning phase.

Box 3-4. Anatomy of an Objective Statement

S —specific
M—measurable
A —attainable
R—results-oriented
T—time-specified

Example: Establish willow woodland on the shoreline of the Rialto reach of Sir Bob Creek by October 30, 2011.

Analysis of the example's objective statement:
S   It identifies the species group of plants; area and location does not identify numbers, but this may be obtainable from design plans. Actions will address the various components of conducting the planned willow woodland.
M   The location of planting is specified; one can reasonably assume the plans will identify species, sizes, and numbers.
A   On the face of it, this appears to be a reasonable task. The reach measurement is not known but can be determined. Questions needing answers include: Do you have control over the planting area? Do you have the resources available to you, the numbers of plants, and so forth?
R   This task is clearly results oriented. This is not an intermediate step.
T   The date is clearly stated. In some cases, a time of completion depends on a prior task; in this situation, stating time in terms of days following the preceding task may be sufficient.

Following are sample goal statements for restoring Cowden Preserve:
1. Increase populations of native wildlife on the Cowden Preserve.
2. Reestablish the native fishery in Kaskaskia Creek on the Cowden Preserve by restoring its stream corridor and watershed.
3. Establish scrubland on the Cowden Preserve capable of supporting populations of targeted upland species.

Referring to the goal statements above, the following objectives can be written for the Cowden Preserve project.

Goal 1: Increase populations of native wildlife on the Cowden Preserve.

Objectives for Goal 1:

1. Establish three acres of mixed cottonwood/willow riparian habitat adjacent to Kaskaskia Creek in the Cowden Preserve by May 2005.
2. Remove all invasive trees and shrubs from the Kaskaskia Creek corridor within the defined project area prior to scheduled dates for the installation of native plants.
3. Eliminate the cowbird on Kaskaskia Creek in the reach from the Elmhurst junction to the Arboga junction by 2008.
4. Salvage mature plants in areas to be graded prior to November 2004.

Goal 2: Reestablish the native fishery in Kaskaskia Creek on Cowden Preserve by restoring its stream corridor and watershed.

Objectives for Goal 2:

1. Remove barriers to fish migration on Kaskaskia Creek by August 2008.
2. Add gravel to two hundred yards of stream bottom substrate to promote fish spawning in Specht Creek (tributary to Kaskaskia Creek) by August 2008.
3. Eliminate stream diversions on Specht Creek by May 1, 2006.
4. Implement erosion control measures on all mapped eroding slopes in the upper Kaskaskia Creek watershed by September 2008.

Actions are the measures that will be undertaken to achieve your stated objectives. Actions are discrete tasks that can be used in the list of events included in the scheduling of your project (chapter 2). An action should be directly related to an objective and should be specific and feasible in terms of resources available (Margoluis and Salafsky 1998). Generally, it takes several actions to accomplish an objective. The more complex an objective, the more actions are required to achieve that objective. Proposed actions may include any site improvements, construction of fencing, removal of vegetation, installation of plant materials, obtaining permits, or organizing another aspect of the overall project that leads to the objective. Actions implemented to achieve an objective can be monitored and evaluated for completion or meeting the schedule.

For the Cowden Preserve example, the following action statements could be generated to implement the above objectives.

Goal 1, Objective 1: Establish three acres of mixed cottonwood/willow riparian habitat adjacent to Kaskaskia Creek in the Cowden Preserve by May 2005. Some actions include the following:

- Purchase the trees (three hundred cottonwoods) from a native plant nursery.
- Purchase irrigation lines, controllers, and so forth, and install them using volunteers.
- Hire personnel.
- Eliminate the weeds.
- Haul debris away from the site.
- Install an irrigation system.

- Hire a contractor to install protective fencing (a) around the site and (b) for specialized protection for specimen plants existing on-site.
- Prepare the site for planting.
- Install three hundred cottonwood trees on three acres of floodplain.
- Collect woody stem cuttings of willow species from other reaches of Kaskaskia Creek.
- Install willow cuttings at a density of two cuttings per linear yard on the lower bank.

Goal 2, Objective 1: Remove barriers to fish migration on Kaskaskia Creek by August 2008. Some actions include the following:

- Map the barriers.
- Determine the equipment needed.
- Develop a plan for removal of barriers at each site.
- Evaluate downstream potential impacts and develop mitigations, if necessary.
- Determine and design temporary access to each barrier site.
- Rent the equipment needed.
- Organize the volunteers and provide safety training.
- Disassemble the barrier.
- Dispose of the barrier debris to an acceptable receiving site.
- Recontour the slopes upstream and downstream of each barrier site.
- Install biotechnical bank/slope stabilization measures.
- Plant native vegetation on slopes in conjunction with the soil stabilization measures.
- Maintain the plantings.

Establishing Project Requirements

Project requirements provide a detailed road map for the project team to follow throughout the project development process. Project requirements summarize and clearly express all the site needs, stakeholder expectations, and imposed requirements for the project. A successful project is one that satisfies all of the known requirements. Project requirements are the synthesis of the goals and objectives for the project and the list of anticipated actions; in addition, they include all environmental permit and agreement compliance requirements.

Translating these inputs into a set of clear and specific project requirements is the next step in the project planning process. This is truly one of the most challenging aspects of project management. It is here that trade-offs are made, and all that is known about what the project is to accomplish is set forth into actionable items. The project requirements serve as the first major milestone for the project restoration team and can be considered a recipe for success. At the end of the project, if the team can demonstrate that each requirement has been fulfilled, the project will be considered complete and will have achieved its objectives.

Project requirements should be documented and published for the entire project team to refer to throughout the formulation of the project. The requirements become a key checklist used to monitor and evaluate team and project performance. Project requirements are the "what" of the project. They describe each facet of the project and what will be delivered to the project sponsor at completion.

Guidelines for Establishing Project Requirements

Project requirements should do the following:
- Be results oriented.
- State the purpose clearly and concisely.
- Use action statement leads, such as *create, develop, monitor, establish, install,* and *remove.*
- Include known milestones (deadlines), such as start dates, end dates, and mandatory finish dates.
- Include illustrations, such as drawings, tables, and charts, wherever possible.
- Obtain stakeholder approval (have stakeholders sign off on the list of requirements).
- Be realistic.

Writing the Project Scope Statement

A basic scope statement includes statements regarding the project's goals and objectives, requirements, schedule, and budget as well as any assumptions made by the project team. Additional information can be included depending on the circumstances of the project in context of the public and working relationships among the project team. Such items can include a signature sheet to record approvals by stakeholders, a list of changes to the project requirements, or a copy of the risk assessment and risk management plan (chapter 2).

PART II

Project Design

We don't often think of ecological restoration in terms of design because many projects do not require much in the way of design. However, for projects on fallow, abandoned, or otherwise severely altered sites, you will have to make decisions on "what goes where." Those decisions are what we call design, whether they are about environmental factors controlling that decision or about the desires of your stakeholders. The most common source for design is to "copy" something already existing, as in a reference site.

The condition of your project site will determine what items will require replacing or reestablishing, and in what shape, pattern, or location on the site. Commonly, this may require you to conduct your own investigations to understand more fully the nature of the various elements in question. In chapter 4, we discuss the various abiotic and biotic conditions that you will need to assess to understand the opportunities and constraints on your project site. If you are not restricted to a specific project site, then you can conduct similar analyses of several possible project sites prior to making your final site selection.

Project design is directly affected by several factors. The final outcome is a reflection of budget, time schedule, and the availability of labor, plant material, and equipment. Adjustments and "rethinking" are typically required. At this point in the process, your creativity and innovativeness are your biggest allies.

In the planning process, you will need to address the sustainability of the improvements you propose to put on the site. Circumstances will determine whether there is a need to provide supplemental water to the site to ensure survival of the plantings, if that was your strategy of restoration. In many parts of the world, water is not as predictable as it once was, or as available. Therefore, you will need to develop your thoughts on how you will get water to the site.

Ecological restoration can involve an extremely wide range of elements. Depending on the project site, you may be able to use several actions to accomplish your project goals. Today, many projects require planting, as there is little time, vulnerable soil, or some other concern that demands stabilizing the ground. Plant material has many forms—seed, cuttings, containers—with some being more appropriate than others. Knowing your options and the requirements for your selection will also affect your budget.

Chapter 4

Site Analysis

In the previous chapters, we introduced a systematic approach to project planning that establishes a framework for undertaking restoration activities, and we began to answer the question, "What makes a project successful?" We found that the first part of the answer lies in the project management process: establishing a method to clearly identify and then achieve project goals and objectives which ensure that specific project strategies adequately and accurately respond to stakeholder needs, requirements, and expectations.

The second part of the answer is using a site analysis process. Figure 4-1 illustrates the many factors (divided into four groups) that should be evaluated during this phase of formulating the project. These factors represent a compilation of several dozen projects throughout North America. The result is a list of factors significantly more elaborate than any one project. Each of these topics is discussed later in this chapter.

We have presented the site analysis process as a wheel because there is no absolute beginning point or order for addressing these factors. Specific site conditions will typically guide you to the most obvious issues. Regardless of the apparent lack of presence on-site, we encourage you to at least touch on each topic, even if it is only to confirm that no issue exists. In doing this, you may avoid unnecessary errors on your project. From the center wheel of figure 4-1, each factor having a response is assigned one of five categories (fig. 4-2) that will contribute to developing a list of project requirements. The center wheel is the SWOT-C analysis (discussed later in this chapter). The responses may lead you to reevaluate your goals or include new elements. The outcome of conducting the SWOT-C analysis is a series of suitable and feasible objectives that support the final goals of your project. These changes contribute directly to the final project requirements.

SWOT-C site analysis begins with data collection, involves data interpretation and analysis throughout the process, and concludes with the synthesis of data that are keyed to your project goals and objectives. Analysis of the strengths, weaknesses, opportunities, threats, and constraints (SWOT-C) provides the context for identifying actionable items that require a response, leading to the creation of conceptual project plans (chapter 6). By assembling a clear picture of the strengths,

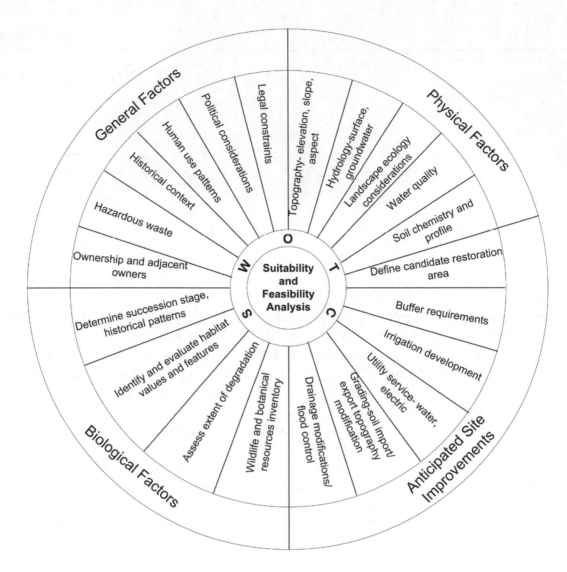

FIGURE 4-1. Site analysis factors by category used in the feasibility analysis process.

weaknesses, opportunities, threats, and constraints facing a particular site, it is possible to antici-pate the necessary site improvements. A comprehensive site analysis will ensure that your site is suitable for the targeted ecosystem and restoration strategies you seek to use in your restoration project. The design approach and conceptual plan begins to take shape as the data are organized into the SWOT-C format.

Pulling the entire restoration plan and design together is no simple task. When the site analysis is conducted within this SWOT-C protocol, you begin to isolate the causes of your site problems and develop appropriate strategies. The intensity of this analysis also depends on the complexity of your site and the extent of the restoration activities being contemplated.

Evidence of detailed site analysis appeared during the development of Hadrian's Villa, at Tivoli near Rome, around 117 BC. Evidence shows that climatic factors, such as wind direction and solar access, along with such site factors as orientation, views, and slope aspect were all considered by the emperor in his design of the Villa complex as a means to maximize the enjoyment of the spaces and the surrounding natural environment (Moore 1960).

More than 2,200 years later, the site analysis process underwent a major advancement when landscape architect Ian McHarg, at the University of Pennsylvania, introduced his graphical method for interpreting multiple environmental factors and their influence in decision-making processes (McHarg 1969). McHarg's method was the first time environmental designers could assemble key site data elements in a uniform format that allowed for easy interpretation and decision making. This uniform method or technique has been further refined with the use of geographic information system (GIS) computer software. Several publications introduce restoration project site analysis (Harris, Birch, and Palmer 1996; Zentner 1994; Packard and Mutel 1997; Bradshaw and Chadwick 1980; Anderson and Ohmart 1985; Daigle and Havinga 1996) but do not discuss the process of using the data in developing a restoration plan.

The ideal site analysis process will focus on identifying the factors or elements that have the greatest potential to influence the outcome of your restoration strategies.

Strengths-
a strong attribute or inherent asset present today

Weaknesses-
not able to function properly, deficient

Opportunities-
a good chance for advancement, future oriented

Threats-
a sign of warning, something impending

Constraints-
something that restricts or limits

FIGURE 4-2. Each factor from the site analysis process is assigned a category and is used for site evaluation and refinement of project goals and objectives.

Collecting Project Site Data

Once you have determined the type of data to collect, the time you have to collect it, the number of people and the skills required to collect it, then you are ready for data collection.

Data on the site factors that potentially influence your project site begin with collection through document search, interviews, and site visits with specialists, such as historians, land managers, local farmers, and nearby landowners (box 4-1). The checklist in appendix 5 provides a useful format to enter data from site analysis investigations.

Box 4-1. General Guidelines Regarding the Collection of Data

1. Assemble and review existing documents that describe the general, physical, and biological factors in the region your site is located.
2. Conduct a preliminary site visit to determine whether the conditions in the field confirm or contradict those described in the reports you have read and reviewed.
3. Contact former landowners and adjacent property owners who can inform you of pertinent factors, studies, memoirs, and reports.
4. Contact governmental agency personnel, natural resource specialists, academic faculty, amateur naturalists, and others who are knowledgeable of past and present general, physical, and biological conditions on your site or in the area.
5. Collect additional field data for factors that are most likely to affect the success of your restoration projects (see the discussion of SWOT-C analysis in chapter 4).

Document Search

The collection of data begins with a search of all available records. City, county, municipal, provincial, state, and federal government offices are often the richest sources. Environmental impact studies, wildlife reports, land use plans, and other governmental reports can contain relevant data and are usually available for easy review. The mapping included in these studies can be a great source of information. These data are commonly detailed and reliable enough to enable the project team to develop a baseline data set from which to begin the project planning. With the widespread use of GIS, these data are often available electronically, which may allow you to develop accurate project mapping.

Copies of environmental studies and reports may also be found at university libraries or research stations. University research projects can be another source of data that could help establish a baseline data set regarding the presence of certain wildlife species, habitats, ranges, and so forth.

Whenever possible, obtain a copy of the most recent soil survey that covers the overall project area. In the United States, soil surveys are conducted and published by the US Department of Agriculture (USDA) Natural Resource Conservation Service (NRCS). Although the soil mapping may not be at the scale needed for your project, the soil survey will help you understand the types of soils, soil profiles, and soil properties that you can expect to find on your project site.

Old photographs are another source of data that can aid the restoration practitioner. Family photo albums and other archived historical records of family farms often reveal much about past land uses and management practices that could influence your analysis. For example, while working on an early settler ranch county park (chapter 14), one team member accessed archived records, including photographs that documented the presence of many species no longer present today. Large stands of old sycamores, cottonwoods, and coast live oak, as well as thick masses of mulefat scrub, that had long ago been cleared away by the cattleman that homesteaded the old ranch were clearly visible. These old photographs provided valuable clues about the site's history that enabled the team to make informed choices about the planting location and species composition that may not have been possible without the historical records.

Many government agencies possess collections of old aerial photographs. In the United States, many of the early aerial surveys conducted by the USDA Soil Conservation Service (now NRCS) date back to the late 1930s in the western United States. Analysis of early aerial photographs often shows the distribution of native plant communities that were later removed for agriculture or other purposes. Sometimes it is possible to assemble a sequence of historical aerial photographs that can be rectified to the same scale, revealing the dynamics of change in the landscape through time. This technique is especially valuable for analyzing the restoration potential at stream, wetland, and shoreline restoration project sites. These photographs also help to date the time of land alterations and structures, which may continue to influence site conditions.

Interviews of Knowledgeable Individuals

At times, documented data are difficult or impossible to obtain. Your only preparation for the site analysis may be interviews with knowledgeable individuals. Local historians and historical societies may possess more than stories and lore. Historic photos, diaries, field journals, and travel logbooks may be available. An extensive program of data collection from museums and other sources was carried out for all of San Francisco Bay, which made extensive use of interviews (Grossinger 2001). Further, former landowners and nearby neighbors may possess documents or detailed personal knowledge concerning previous site conditions, features, and occurrences (Edmonds 2001; Anderson 2001; Fogerty 2001; Harris, Birch, and Palmer 1996).

Site Visit

The purpose of the site visit is to validate data previously collected from documents, uncover new data not yet documented or readily available, and begin the interpretation process for understanding the synthesis of your site's characteristics and the restoration strategies that you envision. Unfortunately, many site visits do not produce the quality of information needed because people often don't know how to detect. The most common mistake we have witnessed during data collection on site visits is the omission of routine or obvious physical site characteristics. To remedy this problem, we suggest that every site visit should occur with proper preparation and a site analysis checklist (discussed below).

Preparing for the Site Visit

Common resource data—such as vegetation maps, topographic data, soil maps, wildlife ranges, overhead and underground utilities, property boundaries, and special easements—should all be compiled and reviewed prior to the actual site visit. These maps should also be formatted for easy access during your site visit. Advances in digital hardware have enabled loading these maps on tablets, notebooks, and smartphones for use in the field. We have used a variety of sizes and formats of data and have found maps sized to fit on 11- by 17-inch sheets work best; they allow the display of sufficient detail while providing a compact space for taking notes and sketching potential design or project technique ideas. Whether you work with a large site and choose 1 inch = 100 feet or a small site with 1 inch = 20 feet, we find that this sheet size is the handiest for combining field notations and sketches and transferring them to plan sheets later.

Varied timing of the site analysis can yield different results. A visit timed to coincide with the spring bloom will yield certain information about the presence of many nonnative species that may lead you to design a weed abatement program. On the other hand, a site visit during a rainstorm may allow you to document surface hydrological patterns, which may lead to the development of erosion control measures for responding to critical areas of concern or may even lead to the identification of opportunities. Ideally, you should plan for multiple site visits, staged over several months, to enable the collection of seasonal site data.

Although we emphasize the importance of a comprehensive and structured data collection effort (Bradshaw and Chadwick 1980; Anderson 2001), the actual factors that you choose to examine and the level of data collected must be determined by you and the project restoration team. The constraints you face regarding time, money, human resources (contractors, volunteers, or paid staff), and stakeholder expectations provide a framework for planning your data collection budget. The demands of each site dictate which factors require attention and the necessary amount of data. The most efficient and fail-safe approach is to develop and routinely use a standardized site analysis checklist to avoid overlooking any critical site factors.

Site Analysis Checklist

When you are on-site conducting a site analysis, it is easy to omit investigating some issues unless you have a checklist. When working through a hierarchy of factors from *general* issues (e.g., historical land use, political considerations, and community involvement) to *physical* factors (e.g., surface hydrology, soils, and climate, including microclimates) to *biological* factors (e.g., vegetation, wildlife, and habitat features), we suggest a process that reveals elements that are essential for project success. From the three categories (general, physical, and biological), we suggest a framework for arranging and cataloging the information collected (appendix 5).

General Factors

General factors are primarily human-originated impacts or conditions that in some way will affect how a project may proceed on the property. Some of these factors cannot be observed on the site but have to be obtained from documentation or other specialized studies (e.g., historical, archeological).

Ownership of the Site and Adjacent Sites. Determine the ownership of the site early to ensure that you possess access rights and to learn about any deed restrictions that may limit or even prevent you from undertaking your restoration efforts. Look for easements (discussed below) that grant to utility companies or adjacent landowners prior access rights across your site. Water, especially on land in the western United States, is not always conveyed with the title for the land. Information about any adjacent landowner's management practices should be considered as a part of your analysis. For example, you may find that annual clearing of fire roads on adjacent lands can lead to surface erosion and sedimentation problems on your site. The lack of an adequate weed abatement program on adjacent properties could result in a perpetual problem with weed growth on your site. Early involvement with nearby landowners and outreach efforts that emphasize mutual understanding and cooperation can result in land management practices on adjacent property that support your goals and objectives.

Easement Rights and Utilities. All easements must be well documented and accurately mapped to ensure adequate consideration is given for the unhindered operation of the utility (e.g., a designated floodway). Also, any long-standing arrangements that may grant access through your site must be documented to ensure that easement rights are not overlooked. In the event that you must maintain access for a neighboring landowner, be sure to recognize this as a permanent feature and to integrate it into your design and management strategies.

A civil engineer or licensed land surveyor is well suited to help you determine all prior rights that others (e.g., utility companies or adjacent owners) may possess on your site. Utility lines (both aboveground and buried)—such as water, sewer, gas, electric, telephone, and cable TV—must be identified early in the planning process so that precautions are taken to avoid utility conflicts. It is best to avoid conflicts with utility services by clearly identifying all known utilities either on or adjacent to your site and planning your efforts to simply avoid interfering with them. Utility companies may provide a pot-holing service that will actually identify underground facilities by digging small holes to uncover and mark pipes and conduits. Other items you should consider documenting include irrigation pipes and drains, culverts, and old footings from service pedestals or conduits.

Hazardous Material. Unfortunately, countless areas possess hazardous waste material that has been disposed of both illegally and legally. Unusual or contrasting texture material should be examined carefully. This is a situation for experts. Essentially, all areas—and particularly remote areas—are candidates to have illegal dumping. Hazardous waste is typically just dumped from trucks, but it can also be buried, backfilled, and the surface contoured to avoid detection.

Historical Context. Archives or contacts with previous owners or neighbors can reveal data related to site features that may affect your design. These may be the presence of an underground spring, a buried foundation or storage tank, or geologic conditions (e.g., a perched water table) that could influence your restoration efforts. Prior land use, both distant and recent, may affect conditions on your site. If your site was used for agriculture, what types of crops were grown? Were soil amendments added? Were pesticides and herbicides used? How well did the crops produce? Could there be deep soil compaction resulting from the use of heavy agricultural equipment?

Current Land Use. Are there existing or proposed land uses on adjacent or nearby properties that might present a conflict or threaten your goals? These will need to be addressed prior to moving on to subsequent steps in your restoration efforts. Otherwise, you will risk making a "fruitless" investment in your project. For instance, conducting restoration activities near a proposed housing development presents a host of issues that must be considered, including vandalism by neighboring children, liability concerns, and free-roaming pets. Neighboring park spaces and open spaces, while potentially leading toward certain synergies, must also be examined to ensure compatibility with your long-term goals.

Political Considerations. Don't neglect the importance of obtaining political consent early in the planning process. Some communities have considered restoration projects to be an attractive nuisance and have sought to block their development. In some cases, the conflicts arise from miscommunication. Conducting public meetings and field trips to properly educate the stakeholders may overcome this issue. Local land use planning agencies can help identify hidden project obstacles and can be excellent facilitators for addressing community concerns.

Access/Access Control/Human Use Patterns. Are there historical footpaths, trails, or roads present on the site? Will an access route need to be maintained? Are off-road vehicles an issue? Is vandalism a concern if a footpath is maintained? Have members of the community used your site as a shortcut to a community center or other features, such as a stream for fishing or hiking trails? Should this access be preserved, prohibited, or controlled? If informal public access has occurred over a long period of time, it may have created a proscriptive easement that will have to be accommodated. These issues must be addressed, and a supporting set of social objectives may need to be implemented in the surrounding community to ensure the long-term success of your project. The need to cultivate a "community involvement" effort is often overlooked by many restoration projects. Left uninformed and uninvolved in the restoration planning process, a surrounding community could unknowingly undermine your efforts. Continued off-road activity, informal recreational uses, unleashed pets, dumping, and so forth can all lead to stress factors that could harm your project's success.

Cultural Resources. Candidate project sites may have historic evidence of use by indigenous peoples or may currently be used by local people for resources or religious purposes. Historical records and university research reports are excellent resources for identifying significant cultural resources that should be preserved or recovered. Whether it is a historic habitation of an indigenous group, a site with religious significance, or "the old oak tree" where previous generations met and carved their initials, you should consider the implications of any proposed restoration activities that might disturb sensitive sites. In most cases, a qualified archaeologist can survey the property, comb the records to uncover any significant resources that could be affected, and offer strategies to avoid conflicts that could delay the implementation of your plan. Contacting local indigenous peoples familiar with your area will identify cultural heritage sites (e.g., traditional sites for the gathering of plant materials, shrine forests, or rock formations) on or near your restoration project site (Fogerty 2001). Some restoration projects (especially meadow and wetland restoration) are being designed to accommodate the continued collection of plant materials (e.g., in certain meadows within the Lake Tahoe Basin) and the use of traditional management techniques such as fire (Lewis 1993).

Agricultural and Other Quarantines. It is not uncommon for pest species to become established in isolated populations throughout the landscape. Many of these species have serious impacts on agricultural crops and have been the target of extensive control measures. Abandoned agricultural fields may have resident populations that have eluded detection until you show up on the site to do a project. Projects requiring excavation or transport with heavy equipment will require inspection to prevent introducing the pest species onto other lands. The local farm adviser or agricultural field office can provide valuable information on any potential concerns.

Stressors on the Existing Ecosystem. During your site visit, you will need to look at the site from different positions to obtain perspective. Evaluate the potential for off-site factors that might present themselves on the project site (box 4-2). Upslope, upstream, or existing roads will be the most common sources for stressors to enter the site. Note the compounding or cumulative effect of some conditions. The lack of vegetation causing erosion may indeed have its origin not in moving or scraping but from contaminants in the soil. Noting irregularities in the occurrence of vegetation is just the beginning of detecting what may have taken place on a project site. Field interviews and historical research may clarify this situation. The occurrence of anything "straight," "square," or "rectangular" may indicate influence by some human-based activity. Evidence of fences no longer present may show the distribution of soil compaction left by cattle; water trough locations may also have experienced excessive compaction. The list can be quite lengthy, and the source of any degradation may be simple or complex.

Box 4-2. Lesson Learned: Have a Watershed Perspective

An agency restored a lowland floodplain forest adjacent to a creek. Although the riparian plantings survived and grew rapidly, the project site was soon invaded by a nonnative species (*Arundo donax*) brought in by floodwaters. The agency was then forced to assume the added ongoing expense associated with the control of invasive plants on the project site. Although the stream reach in which the restoration project was located was free of these invasive plants, many upstream reaches were overgrown with this nonnative, thus providing a continuous supply of propagules for colonization of the project site.

The project planners learned that it is important to understand the perturbations occurring upstream of the project site that could adversely affect the success of a stream corridor restoration project. The agency learned that it may have to institute a watershed-wide invasive species management program to ensure the integrity of present and future restoration projects in the stream corridor.

PHYSICAL FACTORS

Physical factors encompass the structure, composition, and activity of matter, including water aboveground and below ground. Landscape ecology attributes are also discussed because this discipline is concerned with the combined effect of the biotic and abiotic.

Topography. Is there evidence of recent landform changes? Do slope areas present any limitations to plant groups or species? Will access be affected? Sandy slopes greater than 10:1 and other soil types that are 2:1 or steeper present conditions that are ripe for surface erosion, and some regime must be developed in the first phase of your plans to address it. Steep slopes, especially manufactured slopes, are sometimes referred to as "critical areas of concern" and could undermine the success of your project if not modified through various mechanical techniques, such as ripping.

Elevation. Are there freeze lines or frost pockets that will affect the growth rate or growing habit of certain target species? Is the elevation beyond that known for the species being considered for the site? Canyons and ravines typically support species that thrive in moister and cooler conditions. Microtopographic variations may be a consideration when working in marshes and tidelands. Subtle changes of just a few centimeters can affect the success or failure of certain intertidal species. Conventional land surveyors may overlook these minute variations in landform. Make sure that the land surveyor understands your requirements.

Geology. Depth to bedrock, location of rock outcrops, and occurrence of certain formations may restrict using the site as desired. The occurrence of harmful elements or materials could conflict with your restoration or harm your targeted plant species. Some plant communities are restricted to growing only on soils derived from specific geologic formations (e.g., soils derived from serpentine formations).

Soils. You should determine the depth and texture of the soils at various locations throughout the site. Soil texture affects permeability, root penetration, water and nutrient holding capacity, shrink swell potential, and erodibility. Before conducting fieldwork, locate soil maps and reports that may have been made by either the NRCS (previously known as Soil Conservation Service), a local agricultural agency, or the previous or current landowner. Soil maps will frequently provide the information needed to make decisions. The reports containing these maps provide extensive information that may have significant value in your planning process. Information frequently found in these reports includes soil texture, general description of the soil, typical depth from surface, types of vegetation naturally occurring on soils, drainage characteristics, suitability for specific agricultural crops, depth to hard rock or hardpan, and erodibility. Locations in your region are identified with specific data on permeability and available water capacity of the soil as well as pH. Keep in mind that these data are intended to give a general idea of the soils in your area.

March and Smith (2011) used ecological site data associated with NRCS digital soil type maps in conjunction with elevation data to create potential vegetation assemblage maps. These maps were used to guide the selection of the appropriate restoration target. If you lack soil maps, we strongly recommend sampling to assess the differences in soil material throughout your site. Always conduct a field site investigation to check the maps with the site to verify that the landform has not been modified. Methodically plot these locations on a scaled plot map. Identify any locations where the soil appears degraded, whether by salts, chemicals, oils, or so forth. Obtain soil samples near water bodies and from any locations previously identified as areas of concern, as well as wherever you notice changes in vegetation structure or elevation.

How many samples should you collect? Here, common sense and input from the information already collected will guide your decision. If you know that some areas represent disturbed or modified topography, more extensive sampling in those areas may be needed. Areas with no apparent modifications would receive less intense study. If soil maps generated by the government or previous landowners indicate simple soil assemblage of one to a few types, then intensive sampling may not be warranted. Most soil surveys describe the typical soil profile. Dig a couple of shallow test pits to confirm that your site has conditions similar to those described in the soil survey. The presence of certain indicator plant species can provide valuable clues as to the nature and extent of soil conditions. If you have any suspicions that soil conditions may vary significantly with depth, then dig several deep soil test pits to inspect the profile of the subsoil. This action is especially important if you are relying on having your plants tap into the groundwater table. Old fields may have been built up over time, and the subsoil structure may be unknown or dramatically different from the surface. If changing the elevation of the surface is being seriously considered, then collect soil depth data at the time of excavation. Evaluate the depth of sampling to ensure you obtain a clear picture of the soil of your final grade elevation.

Through simple and relatively inexpensive tests, soil laboratories can provide significant insights and offer expert advice for techniques to improve soil structure and plant viability. One cautionary note, however: labs commonly provide information based on crop production rather than native vegetation. References or local practitioners may be able to provide some assistance. At the very least, be sure to have lab tests conducted for pH, salinity (EC), percentage of organic matter content, texture, macronutrients and micronutrients, and any other chemicals that are known locally to be of concern.

Leaving the surface at its present elevation also has its challenges. Abandoned fields and recently used agricultural fields typically are nutrient rich from fertilizers and air quality fallout and thus will promote weed growth over native species. Certain techniques, such as herbicide, irrigation, and dry cycling, can manage the soil seed bank with varying results. One technique that has been recently promoted in Great Britain and Denmark is soil inversion, in which the top three feet of soil is flipped. This technique buries the weed seed bank below the subsoil, which has less nutrients and permits native growth over weed species that thrive in nutrient-rich soils.

Hydrology. Are streams present? What are the flow rates? Is there a nearby stream gauge from which you can construct a hydrograph for the portion of the stream that runs through your site? Do portions, or does all, of your site flood? If data are not available, can you determine the typical high-flow level (ordinary high-water line) by inspecting the banks or adjacent rocks and trees for signs? Often, exploring off-site areas will be necessary to understand what is contributing to drainages on your site. Changes in land use upslope may have an effect on your project site, so try to determine whether there are plans for changes that would increase runoff, such as residential or commercial development. Natural watercourses may be present on your site and can be readily identified. However, if the site has been in agricultural production for some time, there may be structures and landform changes installed to redirect the water to protect the fields. Careful examination of the property perimeter upslope to locate signs of changes in water handling may provide important results. You will need to decide how to address these features. Will changing the watercourses back to their original path be compatible with your objectives?

If you suspect that flooding is a potential issue regarding project success, consult with locals regarding the history of flooding on your project site. The occurrence frequency, duration, and depth of floodwaters will affect your project. You should also consult a hydrologist who can help you determine the anticipated frequency, depth, and duration of flooding for your site based on an evaluation of historic data and the hydrology of the watershed (Kondolf 1995). Dams occur on numerous watercourses, large and small. Their impact on downstream lands is significant and often unpredictable because water releases may not be regulated or consistent. Reliance on water from controlled watercourses should carefully consider the operations of the water regulators upstream.

Groundwater. Determining the depth to groundwater is particularly important when considering wetland and riparian habitats. Installing a series of water table measuring tubes (groundwater observation wells) will enable accurate monitoring of shallow groundwater depths. Depending on your location in the watershed, groundwater can fluctuate dramatically. Seasonal changes should be determined because such information might affect how or when you plant to avoid using supplemental irrigation, or at least determine how long you will require supplemental irrigation. Understanding the fluctuations of the groundwater will determine the types and densities of suitable plants. In some situations, the groundwater is the result of flood events and the groundwater normally fluctuates seasonally, with the plants able to drop roots quickly and follow the descending water table. Often, there is not enough time before starting a project to collect the necessary data. A survey of existing water wells in the vicinity may shed some information on the water levels and water quality.

Surface Runoff. Examine surface runoff patterns to determine the need for erosion control measures. Correcting erosion problems typically requires grading. If your design concept includes earthwork, then you can evaluate various construction approaches to fix the problem; however, for less aggressive projects, there are other techniques to consider, such as placement of rock, wattling, or the use of plantings that can slow the water flow. Surface runoff may be unnaturally concentrated as a result of past activities. This may have affected the vegetation by changing species composition, growth, vigor, and other microenvironment elements. Erosion problems commonly can be handled successfully with vegetation and fiber blankets, coir (coconut fiber) or rice straw rolls (wattles), or mesh fabrics typically nailed or stapled in place, or with various mulches held in place with emulsifying liquids. Despite the recent advances in bioengineering material technology, not all situations can be solved by using vegetation alone. Particularly notable are those situations where water is concentrated, either naturally or through drainage modifications, and the water velocity or frequency is too high for plants to become established.

Water Quality. Water sampling on streams fed by urban runoff is essential to detect pollutants coming from surrounding drainage areas. The levels may be sufficiently high to require some type of treatment. A couple of options are immediately apparent: (1) consider designing a small wetland to improve water quality (bioswale or retention basin) or (2) develop a program of educating the landowners on how to control harmful chemicals and prevent them from entering runoff. The former technique is becoming more widely used as the need for keeping waters clean gains worldwide acceptance. The purpose of a restoration project may be simply to improve the water quality of a stream, river, or lake because the changes in water chemistry will permit the reestablishment of animals and plants previously eliminated from the body of water. A well-organized volunteer group conducting water quality monitoring on the Delaware River, in northeastern United States, has documented significant improvements in water quality. Government organizations are now reestablishing fish populations previously unable to survive.

Landscape Ecology Considerations. In the past few decades, attention has been drawn to landscape ecology, the evaluation of how landscape pattern influences species within various ecosystems and communities. One aspect of landscape ecology receiving attention is biological corridors, core areas, and refuges (Adams and Dove 1989). Numerous habitat conservation planning efforts are addressing these issues (Smith and Hellmund 2006). Corridors are not always nice linear belts of vegetation or habitat leading to large core areas. Depending on the species and communities, the width and quality of a corridor can be highly variable. Just how wide a corridor needs to be is elusive. What is required for butterfly populations to successfully find breeding and foraging habitat, for example, may be unacceptable for small mammals or even some bird species. Understanding habitat requirements of different species and establishing your goals will direct the project design. Commonly, it is possible to provide the habitat requirements for numerous species at the same time.

Another important landscape ecology factor to consider is edge. The term *edge* refers to the change in habitat or structure within a vegetation community. Most often, edge is the linear border where two different vegetation communities meet. However, edge is also present where different age classes meet or where a physical element breaks up the vegetation or creates a change in the vegetation.

Numerous edge patterns are possible, from a very straight, smooth contact edge between a forest and a meadow to an extremely sinuous or meandering contact line between a river and the adjacent riparian habitat. Sometimes an edge is less controlled and may vary depending on the season or cycle of rainfall. While meeting some aesthetic or specific species needs, a longer edge may be more harmful to one or more of the vegetation communities by providing increased avenues for invasion by undesirable species. Research has identified specific tolerances for some species in their ability to invade various plant communities. Observation can show that many exotic or undesirable species may reside in what appears to be a native community. Upon closer examination, however, one can observe several exotic species occupying those small edges or pockets of openings where the competition for space or nutrients is less. Typically, bare or open areas exist in natural vegetation communities, however small. These small openings may become invaded by undesirable species.

This condition will be important when establishing a restored project site. It may not be possible to have a corridor to each core or population center. Your restoration project may be rather small and not connected physically to adjacent natural lands. This should be considered when establishing goals. Detached pockets of habitat will most likely mean that some animals and even some plants may not be able to use the site to its fullest (Morrison 2009). There is nothing wrong with that situation, as long as it is acknowledged and understood that not all possible species will become permanent residents of the project site.

Despite the debate over isolated habitat refuges, it is our opinion that all these sites have value. A series of refuges may be the only means of connecting two core areas that permits exchange of plant and animal genes. The basic question relating to the size of a site is whether it can serve the target species. In many cases, the site will be adjacent to existing vegetation and the combination of the two areas will be sufficient. Knowing the target species' habitat requirements is important in determining what will be needed.

BIOLOGICAL FACTORS

The level of investigations varies with the anticipated changes being proposed and the presence of existing resources that may either be negatively affected or serve as a source for design development. Describing the various attributes of the biotic environment and the individual key species will be important in understanding how the site is being used by those resources.

Existing Vegetation Communities. Taking note of existing native vegetation on-site begins the process of compiling a species list and identifying the communities to be restored. Healthy stands of desirable species can be preserved and enhanced and can serve as sources for seeds, cuttings, and transplanting. Equally important is the identification of noxious, invasive weeds and other undesirable species that may require removal or ongoing management. Decisions on the composition of plant species will strongly influence the ultimate design and management of the project site.

Some sites may have been totally cleared or manipulated, resulting in a complete loss of native vegetation. Information on the vegetation that once occupied the site will require observing adjacent lands or going further off-site. In some cases, records, diaries, and journals from early settlers, biologists, and museum collections can also be used to reconstruct the vegetation community.

Vegetation Dynamics. Regardless of the past degradation of a site, vegetation undergoes constant changes in response to environmental stimuli, such as rain, erosion, soil conditions, and the season. Another variable is the natural competition among plant species.

Vegetation communities undergo a maturation process, with more and more biomass being developed as well as a constant change in microenvironments. Microenvironments may include the canopy of taller plants or the leaf litter depth, indicating a long-term presence of some types of vegetation. Natural processes, such as floods and fire, as well as migrating grazing animals contribute to the physical appearance of the vegetation. If one or more of these naturally occurring processes is absent, then it is likely that the vegetation community will not be on the same track as vegetation in locations with those processes. For example, heath in Great Britain represents an intermediate succession stage and requires active management for its continued existence. Continual human-caused fires suppressed the establishment of pines that normally would have eventually occupied the landscape.

The same situation is found for several endangered species where the vegetation is no longer being subjected to natural disruptive occurrences. In the western United States, the endangered least Bell's vireo occurs in an intermediate-growth form of willow riparian habitat. Periodic flooding of the rivers restores or returns stretches of the river habitat back to early riparian stages. The vireo can relocate to take advantage of these locations as they develop to a stature the vireo finds suitable. Loss of flooding by controlling floodwater with dams has been a significant factor in the loss of succession diversity on rivers in the western United States. Another example is the endangered Stephen's kangaroo rat, which occurs in semiopen scrubland habitat that used to be kept open by browsing deer and antelope, now largely absent from the rat's range. The senior author once recommended a mitigation of maintaining cattle on a portion of a large development to maintain a vegetative state suitable for the continued presence of the Stephen's kangaroo rat.

The underlying principle to remember is that vegetation goes through a series of developmental stages and maturation and then, through some event, is returned to an earlier stage of development. Fire and other methods are needed to replace the original effects of the grazing animals. A site that may not be exposed to significant landscape-forming processes needs to be evaluated to determine whether suitable replacement activity can be provided (Savory 1998).

Assess the Degree of Degradation. The very nature of ecological restoration is to manipulate some parcel of land that is not fully functioning or is missing elements. Degradation activity may be the result of a simple initial impact left to manifest itself into a more dramatic condition by other forces, such as the track left in the desert by an off-road vehicle. Most of the time, site degradation is very straightforward, with only human activity being responsible.

You will have to determine which impact deserves immediate attention and which can be delayed. Experts in the various fields of science may be called on to assist, but most likely it will fall on you and a small team of individuals with varying skills and knowledge to make this determination. Learning to read the land is a talent that requires time and patience.

Invasive Alien Species. Invasive species have become an increasing problem for restoration projects with the increase in the number and range of invasive alien species. The existence of certain invasive alien species on or adjacent to your site may jeopardize the success of your restoration program. Noxious weeds may crowd out or shade out your plantings or compete with your plantings for soil moisture. Some weeds produce substances that deter the growth of other plants.

Habitat Values and Features. Often misunderstood or incorrectly applied, the term *habitat* refers to the sum of the resources and conditions present in an area that produce occupancy by an organism. There are numerous attributes of a habitat that are goals for restoration projects. It is important to clearly identify what specific habitat value(s) or attribute(s) you are intending to establish (Morrison 2009). Most common is the creation of foraging habitat. Although not clearly defined, the implication is typically that the vegetation presence will permit the prey of forage species to take up residence. Insects are the dominant group of prey species in terrestrial habitats. Commonly, there are other components that should be addressed, such as birds needing roost sites, singing sites, and nesting locations. Reptiles, especially lizards, need rocky outcropping or logs where there is an elevation change that will provide suitable sites for courtship display. The absence of these features may restrict the use of the site.

A common project goal is to provide habitat for a species requiring a mature physical feature. Many creative ways exist to introduce these specific habitat features early on a site (chapter 6). Time is important when addressing wildlife values. You will have to decide the habitat values to be included for your site that are not dependent on vegetation. A significant contributor to these decisions will be the goals and objectives of your project. Choosing which values that will be appropriate for your site will largely depend on the habitat requirements of your targeted species. Smaller projects will be more limited in the range of service they can provide.

Wildlife Resources. You should document the populations of wildlife that use the project site. Some species may frequent the site only during migration; others, daily; and still others, only at night.

Compiling an exhaustive list is not usually needed for common or wide-ranging species; however, you should prepare a list of target species and species of special concern relative to project objectives. If adjacent lands have any special status species, then a focused survey should be done for those species to verify use. A natural objective of the project would be to have these species of special status become residents on the site. Commonly, regional lists, environmental documents for nearby projects, and local museums will have lists and other documentation regarding species and habitat preferences.

Obtain data to document nesting, foraging, and use patterns for all species of interest for restoration. As discussed earlier, animals use a wide variety of physical features in their natural behaviors. Some of these features may be critical for their long-term presence (Maehr, Hoctor, and Harris 2001). Studies of the endangered least Bell's vireo (California), for example, show a common pattern for nesting sites to be at about three feet above the ground in a shrubby plant and at a vegetation break or edge with a twenty-foot-high roost site nearby. Other studies have shown a distribution relationship between lizard size and soil surface condition. The ability to lay eggs at the appropriate depth is controlled by the depth of sand and grain size. Larger species require a deeper depth of penetrable sand to lay eggs. If your goal is to increase or establish a permanent population of a species, then knowing the habitat requirements of the target species is absolutely critical to success.

ANTICIPATED SITE IMPROVEMENTS OR CONSIDERATIONS

Almost certainly, a site will show obvious signs of disturbance that will require correction. Determining the cause of these disturbances ideally will ultimately lead to identifying the sources of the degradation and their possible solutions. These potential actions, their extent, and other related characteristics will be important components in developing and refining the project design.

Grading. Past activity on the site may have left the contours unacceptable to the goals of your project. Typically, sites that have served as mining or borrow sites for construction are left as large excavated holes. The desire to establish a wetland type of habitat requires lowering the ground level to be more proximate with the groundwater table or ensuring the appropriate relationship between land surface and tidal elevations. Stream banks may be eroding and have unstable slopes that require modification before bank stabilization can occur. Does the site fit into the vision that you and your stakeholders have for it? Is a significant amount of grading needed, or will you need to bring in dirt to raise the elevation?

Soil Import/Export. Mined lands and borrow sites and other dramatically graded land will be devoid of soil nutrients suitable to support a native vegetation community. The decision to import soil will be contingent on the ultimate vegetation community and the requirements of the species comprising the vegetation. Importing or simply moving soil from one location to another can be very costly. In addition, not properly incorporating the imported soil can leave the site vulnerable to higher-than-normal erosive events.

Water/Drainage. Knowing the source of water on the site and how it behaves will guide you in your ultimate design or planting pattern. The existing groundwater is only one source of water; rainfall and how the water traverses the site will also factor into your design decisions.

All sites have some form of drainage, with varying degrees of slope. Even flat land is not entirely flat and therefore sheds water with small rivulets. Knowing how water travels across your site will be critical in evaluating the need for any erosion control structures or applications. The speed of water movement is a major factor in anticipating erosive events. Some site flooding and planning for those events will greatly enhance the performance of your site. If your site is large enough, you can provide small canals and stream channels that can provide initial water control. The type of flooding will also be a factor in determining the need for protection of banks or levees. In urban areas, water may enter the site from a culvert or a small pipe. Typically, this form of transport will increase the flow rate and the energy with which the soil will move. Providing a rock apron or some other device will dissipate the energy prior to contacting the soil and causing erosion.

Irrigation System. Sites without an apparent water source and with an unpredictable rainfall pattern may require some form of irrigation system to ensure that plants survive. The type of system you use is up to your specific circumstances. If you believe you will need to bring water on-site, you should evaluate the current sources for water, such as an adjacent river, pond, or water pipeline. The source and how you will use the water will be a major factor in your budgeting and design.

Buffer Requirements. It is always a tempting prospect to use all the space that is available to you. Depending on adjacent land uses, however, this may not be wise to do. In urban locations, the adjacent land use is typically a road, a residence, or some other actively used property, such as a park. The immediate area adjacent to your property boundary may be best used by creating some type of screen or natural barrier that will reduce the impacts to the interior of the project site. The requirements of your target species will guide you in determining the most effective screening choice.

Access and Access Control. At times, you do not have a choice on how people access your property. Past land use, agreements, and easements may have already determined that issue. What is at least partially under control is how they traverse your project and the adjacent areas. The more roads and trails across the larger area, the more diminished will be the habitat values. Access control is a critical issue, especially with young sites. A wide range of fences and other barriers can be used very effectively. Will this be an issue for your project? Is your project "in the way" for local residents trying to get somewhere else? Sometimes it is more productive to have a designated path or road that controls where the visitors will be rather than trying to prevent any access.

Vandalism Control Features. Unfortunately, vandalism is a significant issue with urban restoration projects. The reasons for vandalism are many (chapter 10) and may be beyond your capability to eliminate entirely. The best you can do is design a project that addresses the types of vandalism you anticipate as reflected in the adjacent land uses and human pattern of the vicinity. Most often, irrigation systems are attacked. We have had our sites used for practicing baseball batting, making off-road vehicle donut tracks, and trying out a new hatchet, to name a few. How will you address this issue? There are measures to control access by vehicles, but keeping people totally out of a site is generally not feasible. Coming up with an approach to counter vandalism requires creativity (chapter 10).

Define the Candidate Area for Work. Make maps of your site during your visit. Do not rely on engineering maps alone; instead, look at the site from different points of view (fig. 4-4). Mapping will help you come to terms with just how much of the site will be included in a specific project. Several factors will come into the decision-making process for how big the project should be. What shape will it take? Perhaps a portion of the site will require excessive resources, leaving you with limited funds and resources to restore the rest of the site. Your initial thoughts will need to be recorded and reexamined from time to time.

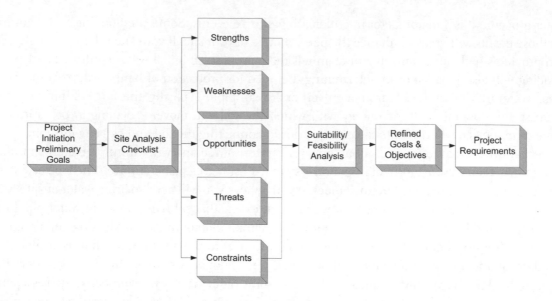

FIGURE 4-3. Information developed from the SWOT-C process will directly contribute to refining preliminary project goals, leading to development of project requirements

Interpreting and Analyzing Site Data

Often, we have seen situations in which, despite site conditions that we considered obvious, project teams have failed in project implementation because of a misdiagnosis of the site conditions. Upon further checking, we have found that the data were indeed collected and analyzed. However, the way the data were interpreted led to a series of restoration strategies that later proved ineffective. Given this, how can we ensure that we properly interpret the factors we uncover in our data collection process?

Using a site analysis process ensures the collection of relevant data, which can then be arranged into five broad categories for interpretation and analysis: strengths, weaknesses, opportunities, threats, and constraints (SWOT-C). The SWOT-C tool is adapted from Philip Kotler's (1999) work on examining strengths, weaknesses, opportunities, and threats in business competitive analysis. The SWOT-C analysis offers the restoration practitioner an effective framework for organizing data collection and the synthesis process (fig. 4-3). Organizing your data collection into such categories creates the basis for a systematic approach to the analysis.

After the collection of the data has been completed, it is time to put the various pieces of information into a format from which you can develop a plan for your restoration project. The SWOT-C process identifies issues that you must address when developing your restoration project.

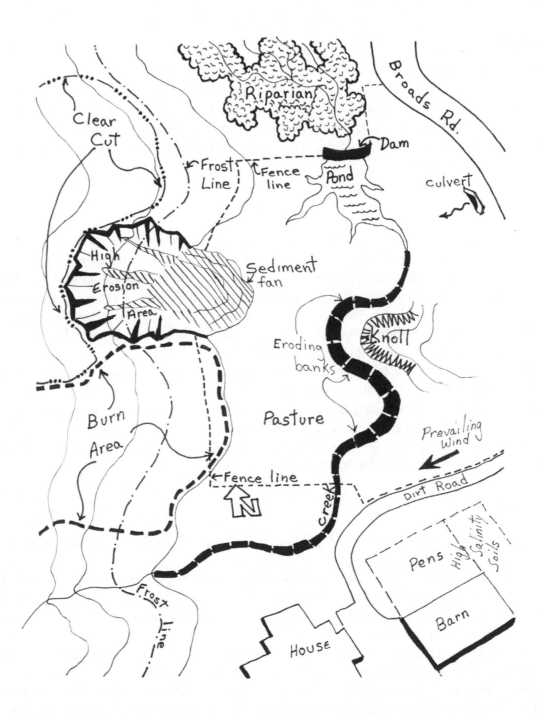

FIGURE 4-4. Compilation of notes and mapping of a project site after several field trips to the site.

Strengths

A strength is *a strong attribute or inherent asset* (Merriam-Webster 2003).

Typically, strengths are factors that positively influence your site and are desirable assets that should be not only preserved but enhanced. A strength is considered a somewhat relative term; on your site, a strength may constitute, for example, a remnant patch of native grasses or an intermittent stream. Such strengths may provide a starting point, or even the main theme, with which to begin your design efforts.

Weaknesses

A weakness is *a fault or defect, deficient in physical vigor* (Merriam-Webster 2003).

Devote your time to gaining a clear understanding of the negative trends or patterns on the site. Look for obvious stressors. Are there patterns of erosion or sedimentation that may require remedial action? Are adjacent land uses supporting or causing apparent weaknesses? Are current management practices of adjacent, upwind, or upstream lands causing degradation on the site? Are invasive weeds populating the site? Develop a list of weaknesses and consider rating them according to their relative importance for remedial action.

We usually break down all issues into two categories: issues requiring immediate attention and issues that do not directly and immediately affect the implementation of the restoration strategies. The list of weaknesses may become a list of issues that are critical first responses because anticipated project features cannot be properly implemented without first addressing the project weaknesses. Depending on the extent and degree of the remaining weaknesses, you may consider identifying anticipated improvements to be implemented in future phases.

Opportunities

An opportunity is *a good chance for advancement or progress; a favorable juncture of circumstances* (Merriam-Webster 2003).

A major purpose of the site analysis is to discern opportunities where restoration can reestablish functions and values. Often, it is the subtle issues that tend to influence your decisions the most. Once, when conducting a site analysis for an oak woodland restoration project, we overlooked the presence of thousands of oak seedlings. The site had been frequently grazed by cattle, and all the seedlings had been browsed to the ground. It wasn't until a subsequent site visit, after the cattle had been moved to another location, that we first noticed hundreds of oak seedlings throughout the area. Consequently, we completely reoriented our approach to the restoration effort, and rather than the intensive planting regime we first considered, we simply fenced the site and kept the cattle out, thus allowing the seedlings to grow.

The above illustration reveals that opportunities may be very well hidden and that it can take several site visits to uncover them. Much of your identification of opportunities depends on your view of the site when considered through the framework of your project goals. Consider urban runoff. At first thought, our reaction is usually tilted toward the negative, and we think of nonpoint source pollutants (road oils, salts, nitrogen, and phosphates), peak flows, erosion, and a host of other problems. And yet, we have seen streambeds, once considered intermittent, flowing not just after peak storm events but year-round and supporting riparian habitats that function as host to many wildlife species. Therefore, be sure to take the time to reflect on all that you uncover during your site visits. Discuss your impressions with other team members; and be sure to consider your data collection in light of the project goals.

Threats

A threat is *an indication of something impending* (Merriam-Webster 2003).

Site analysis should identify where an undesirable trend or challenge may lead to the further degradation of the site. Threats should be classified according to their seriousness and probability of occurring. A risk analysis (chapter 2) should be conducted anytime the probability of occurrence is uncertain and the potential for impact to the project site appears high. The site assessment should begin with identification of the critical areas of concern that pose a direct and real threat to the on-going natural processes on-site. Once identified, the critical areas of concern can be traced back to their causes, and actions must be taken to remedy the problem(s). Depending on the severity of the problem, these critical items are most commonly treated early in the implementation of the project.

Constraints

A constraint is *the state of being checked, restricted, or compelled to avoid or perform some action; a force by imposed stricture, restriction or limitation* (Merriam-Webster 2003).

Project constraints are not limited to environmental factors only. They necessarily include economic resources (e.g., working capital), human resources (skills, number, and availability), time, stakeholder requirements or expectations, seasonality, political considerations, climate, natural resources, and so forth. Constraints such as these require some form of response from the project restoration team. There may be governmental agency restrictions on the seasons when brush clearing or grading can take place. Working in or adjacent to occupied habitat during the breeding season for certain wildlife species commonly requires provisions controlling when and how the work can be performed.

The restoration practitioner should always view the site analysis phase as an attempt to capture a snapshot in time that characterizes the overall trends of the site's condition.

Reflecting on the Data and Your Observations

You may develop impressions of the site that do not necessarily fit into the framework of the SWOT-C analysis. These impressions are nonetheless important to consider in your analysis. The synergy of data and impressions on several factors can permit you to reach a conclusion that otherwise may not have happened. The ultimate goal is to make use of all information you have available, whether tangible or not. The restoration of Carrifran Wildwood is an excellent example of a volunteer-based project that effectively used the SWOT-C process (box 4-3).

Box 4-3. Restoration Project Highlight: A Comprehensive Project Plan Is Important, But It Must Be Flexible Enough to Change as More Information Is Revealed from Detailed Site Analysis

Location: Carrifran Wildwood Woodland Restoration Project, Dumfriesshire, Scotland

The mission of the Carrifran Wildwood Project is to restore an extensive tract of mainly forested wilderness with most of the rich diversity of native species present in the Southern Uplands of Scotland before human activities became dominant. The Moffat Hills area holds a wide range of upland plant communities, including herb-rich flushes and calcareous ledge communities Vegetation types found in the western Moffat Hills include smooth grassland, flush, tall herb, blanket bog, wind-clipped summit heath, dwarf-shrub heath, tussock grassland, and bracken, resulting in the richest assemblage of montane and submontane plant species in the Southern Uplands.

Most of the site has suffered from overgrazing, which appears to have reduced the area of heather (*Calluna vulgaris*) and has effectively halted tree regeneration except on the most inaccessible ledges. The total project area is about 1,600 acres; the total area planted is 740 acres. During the first decade of work, more than half a million trees were planted, establishing young woodland in most of the lower half of the valley. Eventually, this will lead to development of one of the very few extensive areas of treeline woodland and montane scrub in Britain.

The project came about through grassroots efforts because many residents have become increasingly uneasy that their familiar, beautiful, but mainly naked countryside has been ecologically devastated, with only a fraction of the biodiversity that was once present. At present, there is nowhere in the Southern Uplands, and few places in Britain south of the Highlands, where one can get a feel for the natural vegetation of the countryside on a reasonably large scale.

The Wildwood Group achieved the purchase of Carrifran Valley in 2000 (fig. 5-2). The group had forty to forty-five active members with an impressive range of expertise, including professional foresters, ecologists, botanists, and zoologists, as well as people with a high proportion of other professional skills (e.g., law, education, horticulture, information technology, and business management) needed to carry through such an ambitious project. Inspiration and outreach to the wider community were provided by the strong representation of artists in the group.

A management plan describes the goals for woodland restoration at Carrifran. After three years, it represented the fourth main iteration of the document. By late 2000, as the process of restoration had already begun, some minor amendments to the original management plan had already been made—evidence of the adaptive management approach in action. This document represents not the final word on how restoration will be achieved at Carrifran but, rather, a flexible set of blueprints that will need to be adapted to changing circumstances as the project develops.

Carrifran is a small valley, and although it comprises a natural entity, in landscape terms it cannot be isolated ecologically from adjacent areas. Constructive relationships with landowners in the area have been made, in the hope that long-term changes in land use may gradually create "buffer zones" matching with Carrifran where management of plants and animals will be in harmony with the development of the wildwood ecosystem.

The management plan was developed for Carrifran in advance of its purchase. The plan was based on a thorough investigation of several resource elements found in the Moffat vicinity, where Carrifran is located. Physical characteristics examined and discussed included the climate, hydrology, geology, geomorphology, and soils. Biological characteristics included current distribution and status of woodland, vascular plants, vegetation, bryophytes, fungi and lichens, mammals, birds, fish and other vertebrates (amphibians and reptiles), and insects and other invertebrates. The main factors influencing the establishment of woodland on the site are the soils, climate, existing vegetation, and landscape. The planning group identified the constraints and opportunities associated with geology, hydrology, soils, flora and fauna, landscape, archeology, and recreation. These data are serving as an effective tool in the development of projects and further actions in the Carrifran Wildwood.

The level of study and documentation clearly reflects the broad spectrum of professional individuals volunteering for the site analysis. An archaeological study was conducted by Historic Scotland documenting eleven features or groups of features for the property. A major element of interest was added to Carrifran by the discovery on the site of an Early Neolithic flatbow, made of yew and dated at 4040–3640 BC; it had been broken and was presumably discarded by a hunter, after which it lay preserved in the peat until it was found in 1990 by a hill walker. This is the oldest bow ever found in Britain and is now exhibited in the Museum of Scotland in Edinburgh.

Historical and current adjacent land use issues were identified and discussed; most notable since World War II has been the establishment of conifer plantations. This poses a potential threat to Carrifran because the conifer could become established by natural means. Recreation and access issues were also addressed and helped form the basis for the overall management plan.

Extensive research and data are included in the management plan related to the woodland structure. Elevation gradients among the various species comprising the woodland have been documented by others in various parts of the region. This formed the basis of the restoration plantings distribution and the planting compartment configuration. In addition to a detailed planting scheme, the plan also includes a grazing management plan with criteria for phased removal. The plan progressively controls sheep grazing on the site to prevent entry into newly planted areas.

Establishment and maintenance addressed the potential for natural regeneration on-site. Many species have this capability, and with the removal or containment of grazing, these species may be able to establish on-site without direct human involvement. There are limited native species in the vicinity, and the prospects of their being brought on-site by animals is slight for most. Direct seeding is an option that is less costly and can provide plantings in more difficult locations on-site. Sown areas can be fenced as required.

Ground treatment and weed control, nursery operations and considerations, species introductions, and habitat features, such as freshwater habitats and bogs, are also discussed to complete the overall management plan (abstracted from Newton and Ashmole 2010).

Chapter 5

Design Approach

The outcome of developing goals and objectives and conducting a SWOT-C analysis is to develop the project requirements, which—along with a risk assessment—create your project scope, a major element in guiding your project design (fig. 5-1). In this chapter, we discuss four approaches to developing a reference model for designing a restoration project: extant reference site, historical reconstruction, remnant, and fabrication. Ecological restoration projects must have some form of ecological reference for project design and from which comparison and evaluation can be conducted. The use of a reference to develop a reference model is the most common approach for designing a restoration project. Using a reference site incorrectly or not understanding the selected reference site weakens the integrity of your project.

White and Walker (1997) propose four categories for reference sites: (1) same place, same time; (2) different place, same time; (3) same place, different time; and (4) different place, different time. The most common categories are the first and second, when it is possible to visit a site prior to its damage (e.g., mining) or use a nearby site typically adjacent or nearly so to the degraded site.

Selecting a reference site requires the evaluation of several factors and an awareness of the historical context of the site. Ideally, reference sites should have as many of the physical characteristics as possible in common with the project site, which is sometimes a challenging task. Collecting data at several sites might overcome this deficiency. In the final analysis, the task of data collection from a reference site is to assist in the design aspects of the project. The reference model should be considered a guide or beacon (Clewell and Aronson 2013).

Reference sites typically are mature; in some instances, it may be possible to collect data from developing vegetation, which provides very valuable "midcourse" information. Species appear and disappear from some ecosystems over the time it takes for a site to become fully developed. A site will mature while reacting to several environmental stressors affecting the site (e.g., fire, flood, animal activity). Responses to these stressors will change species composition, distribution, and vigor. With sufficient time, additional species may become established as microhabitats are created in addition to the plants growing in size and maturity. Collecting data from sites of varying

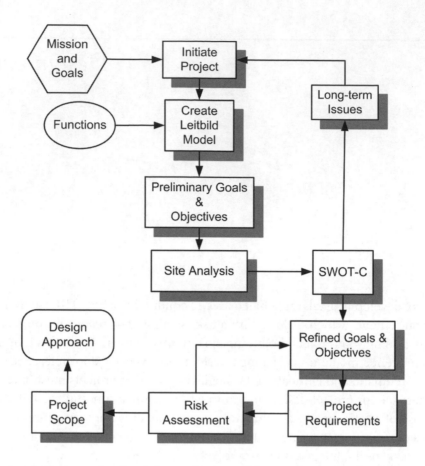

FIGURE 5-1. Project initiation tasks. Careful consideration of these tasks is critical to success.

ages can capture this maturation process for a reference site design model and later as an analytic tool for evaluation of performance. The compiled data will establish a trajectory. Project progress can be evaluated by comparing the trajectories of the site with the reference model.

Appropriate selection and use of reference sites will provide a strong foundation from which to launch your project.

Four Design Approaches

The four design approaches—extant reference site, historical reconstruction, remnant, and fabrication—are all approaches for creating a reference model for your restoration project. These approaches can be used in combination as project site conditions permit. *Extant reference* is the use of an existing ecosystem on-site or nearby with similar physical characteristics. The use of an extant reference site is the most common method of determining the species composition and other characteristics. *Historical reconstruction* relies on literature, photographs, historical documents, and oral history. *Remnant reference* is the result of collecting data from several small, isolated examples and compiling the data to compose an overall description of the ecosystem to create a restoration model. *Fabrication*, or creation, is where the ecosystem did not occur previously but adjacent or nearby ecosystems provide the species composition and physical attributes (Clewell and Aronson 2013). The design approach you choose for your project depends on the goals and objectives you developed early in the planning process.

Extant Reference Site

By comparing the species and patterns on the actual, adjacent, or nearby site, it is possible to design the vegetation and habitat you desire (fig. 5-2). This works fairly well in situations of isolated degraded lands near or adjacent to natural vegetation with limited disturbance. The specific data to collect will be determined by the type of vegetation and habitat under study. Typically, such data will include cover by species, species lists, vegetation pattern, unique habitat features, species densities, and length of edge between different types of habitats or vegetation. Following

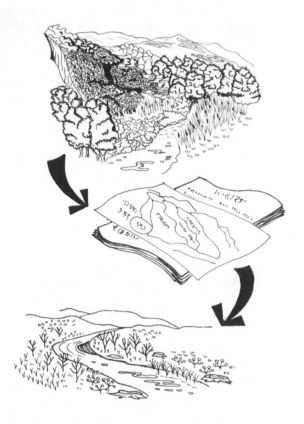

FIGURE 5-2. Use of a reference site will assist in developing a site with similar plant composition in most cases. The objective is not to directly copy but to emulate the functions and general attributes of the reference site.

an initial visit to the proposed site, you will have an idea of the stressors and resources requiring your involvement. If the soil is not damaged, then other than verifying a similar soil type, it will not be necessary to conduct further studies. You should gather data that is needed for recovering impairment.

Data routinely collected by botanists and ecologists do not necessarily lend themselves to providing usable data for the development of a design. For example, knowing that vegetation has .0013 stems per meter square is almost meaningless for design building. Stem counts do not necessarily relate to the number of individual plants. Plant demography is going to sort itself out at a restored site regardless of how picky you are about planting densities. It is more important to ensure that all of the desired species are present and that there is sufficient abundance of key species and life-forms for community structure to reassemble itself promptly. Collecting data for restoration designs does not necessarily conform to typical statistically designed data collection methods. It is often best to develop your own reference site database.

Developing your own reference database has several advantages. By knowing what your goals are, you can collect the data directly related to those goals. Use prior available data to your advantage, if it exists, but be selective and only use that which is relevant for your purposes. This eliminates approximating or interpreting the data of others. The database forms a strong foundation for developing a specific design. However, there are limitations to using reference databases. These limitations involve the elements of time and location.

If your data are collected from a vegetation community with a long life, then you must make sure that the various environmental factors affecting your site are similar to the reference site. For riparian areas, this may not be straightforward, as the dam building programs during the past century have modified the hydrology of numerous rivers and streams (box 5-1). Thus, the riparian vegetation growing on a site may have been established under a different set of hydrologic conditions than those conditions that presently affect your project site.

Box 5-1. Lesson Learned: Choose Your Reference Model Carefully

Restoration project planners used a nearby mature floodplain forest as a reference model for the design of a riparian habitat mitigation project. The planners failed to recognize that the surface and groundwater hydrology of the reference site had changed considerably since the inception of this riparian forest as a result of dam installation, groundwater withdrawal, land development, and water management activities in the watershed. During the permitting process, success criteria were tied to achieving a trajectory of vegetation growth that would lead toward the characteristics of the reference site. Whether it will be possible to replicate the reference site in the long run is questionable due to the changed conditions in the watershed hydrology. Only time will tell.

The project designers learned that it is important to avoid choosing a relict stand of vegetation as a reference site. They also learned that it is better to design a native plant revegetation project based on a hypothetical model that takes into account current watershed conditions.

Another example addresses weather: a series of wetter-than-normal years can influence the species germination and pattern where in normal years the species would be a minor element in the landscape. This situation is difficult to assess if you look at only one location over a short time. A technique to handle this potential bias is to study several sites in the vicinity (Bonham 1989). When working in a stream or creek where the natural hydrologic cycle or intensity has been altered, you can expect that the vegetation being studied was established under a very different hydrologic regime, one that may not occur today, so adjustments will need to be made.

The presence of unusual vegetation communities or species may prompt an investigation to understand the circumstances that account for their presence. A period of unusual rainfall or drought might produce site conditions that enable establishment. After a short time, the conditions may no longer exist, but the species or assemblage may have matured to a point where they can persist. Determining the age of the vegetation or assemblage will enable you to determine the climatic conditions during the time when the vegetation was established and will permit comparison with the present-day climatic conditions. If current climatic conditions are similar, then establishing this assemblage may be feasible. Aging the vegetation can be done in a number of ways, from coring of tree trunks or cutting cross sections of shrubs and analyzing the rings to analyzing a series of old aerial photographs.

In some cases, the primary issue is the soil surface condition and the possible accumulation of chemicals and heavy metals. In many arid-land systems of the world, salts accumulate in the upper soil layers and prevent the natural reproduction of plants. However, because the more mature plants have roots several inches to several feet below the surface, they are unaffected by the buildup at the surface. In these cases, it would not be advisable to plant or sow seed in the currently unvegetated locations because the conditions of the site have obviously been altered.

Another consideration in using a reference database is the distance the reference data was collected from the project site. Even though you may be in the same vegetation community and at similar altitude, the reference database may be considerably different. How far away from your site can you rely on reference site data? Unfortunately, there is no rule of thumb. We have collected data for an endangered bird species in southern California scrubland. Only fifty miles farther north along the coast, the vegetation, composition, and structure of southern California scrubland differs markedly in species composition and cover characteristics. It is always advisable to conduct some initial data collection and compare it with your intended database before committing too many resources to the project. This simple preliminary step can eliminate lost time and resources.

As with almost anything involving sampling, it is best to have a number of locations so that these differences can be merged and variation ranges can be determined and incorporated in the planning and design. A significant project in Scotland has, for the past fourteen years, been restoring an entire watershed using data collected from the adjacent extant stands of native vegetation as well as using library resources to compile their reference model (box 4-2). Figure 5-3 shows Carrifran Valley at the beginning of the planting operations in its first year. After twelve years, essentially all of the areas designated for forest restoration have been completed, with the focus on monitoring and adding the more rare species from the area. In this example, the predominant approach is a reference model; however, they have used historical reconstruction and remnant approaches to further enhance their model and increase the diversity of species reintroduced onto the site.

Historical Reconstruction

The historical reconstruction approach refers to the past condition, frequently targeting a time of some significance to the region or community of the project. Although aspects of this approach include preservation techniques, this approach relies on restoring landscapes that for whatever reason were removed from the site.

Historical parks, monuments, and sites of cultural significance have their models based on the historic, and sometimes the archeological, record from the region (Egan and Howell 2001). Re-

FIGURE 5-3. Once farmland, now a Scots pine forest restoration project (see box 4-2). Carrifran Glen watershed, Moffat, Scotland. (Photo by John Rieger.)

gardless of the specific site, historical reconstruction has a defined target as described from the historical research, and although the details may be lacking, the general goals have been formulated.

Sources of information to describe the historical condition include photographs, diaries, oral histories, maps, and expedition journals. In many instances, knowing the distributional history of plant species can tell you whether they would have been present.

In Israel, the reconstruction of Neot Kedumim, an ancient community from biblical times, has been accomplished by using scriptures as the primary source in defining target communities (Naveh 1989) and integrating many values and uses of the historic ecological community and its plants and animals (box 5-2).

Use of the historical reconstruction approach for developing a reference model need not be geared to a time period several centuries in the past. In many cases, this is not even possible. A regional park comprising natural canyon lands and a historic ranch from the early days of San Diego has as its historical target the period 1862–1872, the time the ranch was in operation (chapter 14). Using numerous records and sketches, and some photographs, of the ranch, we identified a missing component of the riparian system, Fremont's cottonwood (*Populus fremontii*), long absent due to having been eliminated for firewood and water conservation.

Box 5-2. Restoration Project Highlight: Historical Approach for Ecological Restoration Project Design

Location: Neot Kedumim Biblical Landscape Reserve, Israel

In Israel, an especially significant project has been developed, and continues to evolve, that draws on the biblical and Talmudic literature for the selection of species of plants and animals as well as descriptions of the landscapes. Located in the Modi'in region, halfway between Jerusalem and Tel Aviv, Neot Kedumim is more than a garden and learning center. By accepting the interrelationship between man and nature, Neot Kedumim accepts the presence of human-altered landscapes as well as the natural ecosystems that occurred in the past among the agricultural activities. The ancient Mediterranean landscapes have been influenced by human activities for more than half a million years and presumably coevolved together with Paleolithic man since the middle Pleistocene (Naveh and Lieberman 1984).

Covering 550 acres of rocky and denuded hill land, the garden is located on the site of ancient agricultural and grazing pastures used continuously for thousands of years. Extensive degradation had occurred on-site, with massive amounts of erosion to the point where only rock remained. Beginning in the late 1960s, major restoration work transformed this wasteland into landscapes representing those described in the literature. Several ecological communities have been restored, such as the Sharon Forest of the coastal plain and open woodland of tabor oak (*Quercus ithaburensis*) (Naveh 1989).

Restorationists have planted flowering plants occurring in the Song of Songs, which include Sharon tulip (*Tulipa sharonensis*), the sand lily (*Pancratium maritinum*), the Madonna lily (*Lilium candidum*), and narcissus (*Narcissus tazetta*). Using this historical approach to restoration will naturally raise the question of authenticity. In this example especially, there is wide latitude for interpretation. But can anyone decide what the precise assemblage is? Using the interpretations from the literature sources and making informed decisions among various experts seem to be the best approach.

Using the historical reconstruction approach poses several challenges, the most significant being whether an adequate description exists not only for the plant species but also for the physical conditions of the site. A species list alone says only one thing about a vegetation community—namely, what was present. It does not indicate which species were more common than others. A reasonable approach to addressing this situation is simply to make sure that sufficient individuals of all species are present and in various areas to ensure that they will self-sort and establish themselves over time.

Remnant Patch

In some cases, the land to be restored may have a few existing islands of undisturbed natural vegetation that can serve as points of reference. As with the extant reference site approach, one method of overcoming this situation is to collect data and material from a wide range of sites of varying sizes throughout the vicinity (Packard and Mutel 1997). Using this technique will help develop a composite picture of the species possible at your restoration project site. These patches can be created by, for example, the proximity of two or more railroad tracks, older cemeteries in rural areas with little to no maintenance, or isolated strips along roads and other rights-of-way where grazing and human disturbance were prevented. In Asia, isolated shrine and temple grounds have protected vegetation remnants; a similar situation also exists in some villages of India, where sacred forests can provide sources of material.

The presence of a plant species depends on several environmental factors, including the soil type and its condition, elevation, slope, and even when it was last physically disturbed. Most plant species require pollinators to produce viable seed. The ecological requirements of those pollinators need to be met to be present on the remnant. Depending on the time of isolation, there may have been several species that no longer are present. If a remnant was part of a larger community that requires fire periodically, the isolation and cessation of fire may have negatively affected those species needing fire for seed germination. Understanding how the ecosystem responds to an environmental regime will assist in your determination of the appropriate species to include in your reference model.

One of the most dramatic examples of using patch remnants is the prairie restoration effort of the Great Plains of North America, with such projects as at the Fermi National Accelerator Laboratory in Illinois and the Curtis Prairie in Madison, Wisconsin, being notable examples. At the Fermi Lab outside of Chicago, a volunteer group began in 1975 that was to lead to a five-hundred-acre prairie on former farmland inside the accelerator (Nelson 1987). The approach was to visit several small patches of prairie remaining in the general vicinity. Tabulating species and collecting seed from these sites formed the initial plantings of the prairie, which now comprises several hundred acres. Comparison with these remnant patches also indicated the absence of some species at the prairie project. Some species were less common, and seed had not been collected. Using these various data sets, the group set about to collect and propagate those more uncommon species or species with unique habitat requirements and introduce them onto the project site. Approximately forty years earlier (in the 1930s), this same technique had been used on the Curtis Prairie (fig. 5-4), established also on abandoned farmland, this time at the University of Wisconsin Arboretum. The project was a vision of Aldo Leopold and was executed by Theodore Sperry and a team of Civilian Conservation Corps workers. The approach was the same as described for the Fermi Lab.

Fabrication

A very common practice today, and one largely attributed to the "no net loss of wetland" policy, is the establishment of an ecosystem on land that previously did not have this ecosystem—thus, it is

FIGURE 5-4. Example of a remnant approach: the Curtis Prairie restoration project, on an abandoned farm that is now part of the Arboretum of the University of Wisconsin, Madison. Restoration began in 1936 and is the oldest documented prairie restoration project in North America. (Photo by John Rieger.)

not true ecological restoration. This practice is more appropriately called fabrication or creation, and it typically is associated with compensatory mitigation requirements in the environmental permitting process. Fabrication should, however, still employ the various principles and steps that are present in true ecological restoration projects as described in this book.

In many cases, selected sites require extensive changes in topography and elevation to create the hydrologic conditions required of the created ecosystem. Species composition is derived from sampling nearby and may be a composite of several sites. In extreme situations, such as undeveloped lands in the middle of an urban landscape, there may not be any representative ecosystems nearby. In a project conducted in west Los Angeles, for example, the senior author compiled a list of species using various remnants, references in a journal, and herbarium collections from the general region. This approach is typically used in extensively cleared landscapes and where adjacent natural areas are absent. As part of a land use plan for derelict industrial lands, fabrication and ecological restoration has been under way in the Ruhr Valley of Germany. An example (fig. 5-5) can be found at an aluminum factory, using a small reference site, one eighth of an acre, believed to date back to the mid-1800s.

FIGURE 5-5. Once-derelict industrial land in the Ruhr Valley of Germany. This and several other sites now support a diverse assemblage of species. Essentially no native vegetation community remains in this region. This site reflects the fabrication approach. Essen, Germany. (Photo by John Rieger.)

Chapter 6

Design

Restoration design can be a challenging, iterative process requiring the individual to satisfy the project requirements (goals and objectives) while balancing the scope, schedule, and cost constraints of the project. Design is the plan or intention of an action or object prior to its actually being built or executed. We consider any conscious decision to make modifications (e.g., species introduction, reestablishment of fire) as an act of design.

Several tasks contribute to the final project design, and when they are combined, we refer to this portion of the project as design development. We begin by describing a conceptual restoration design. Then we discuss design development, in which the details of the design are worked out and conflicts in the design are resolved prior to the producing the project plans (fig. 6-1).

Developing the Conceptual Design and Concept Plan

The process of translating the project requirements into project plans begins with the development of a schematic diagram of the major project features. Several methods can be used to accomplish part of the design. Several drawing/mapping computer programs can be used, including ArcInfo and CADD. The diagram shown in figure 6-2, which is sometimes referred to as the "'bubble'" diagram because of its bubble-like shapes, is used to depict the key elements of the design. It begins with a scaled plan (or base map) that depicts the existing features on the project site.

In many cases, the best base map to use for the concept design is the site analysis plan. Working in a small group to allow everyone to participate in the design process, we begin by diagramming key project features using a thin overlay of tracing paper on top of the site analysis plan. Whether you use this technique or a computer program, you should allow for making changes of varying magnitude and frequency. We use a broad felt-tipped pen to draw the proposed design on the tracing paper. This technique provides a connection between the field analysis and the project requirements. The bubble shapes represent the key project features that the design is required to contain. For example, in figure 6-2, one bubble represents a riparian area that is to remain. Here,

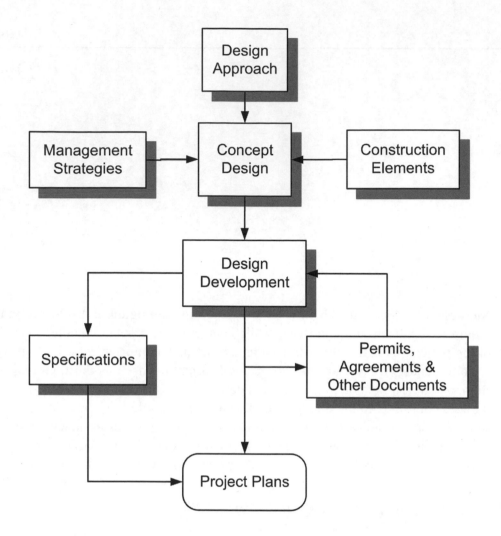

FIGURE 6-1. The design phase uses the project initiation products to develop the approach and look of the restoration project.

you would roughly outline the area that is to be preserved. Then you would place shapes for new riparian areas that must fit in and around the existing riparian habitat. Open meadow areas are depicted with a bubble. Nonnative species that are to be removed are identified with a bubble.

Identify Key Project Features and Requirements

The focus is to place or locate the key project requirements on the diagram. Don't be too concerned with how accurately the shapes represent the actual project features. Buffers can be visual or physical barriers placed to block the view or to discourage access into an area. Perhaps at the

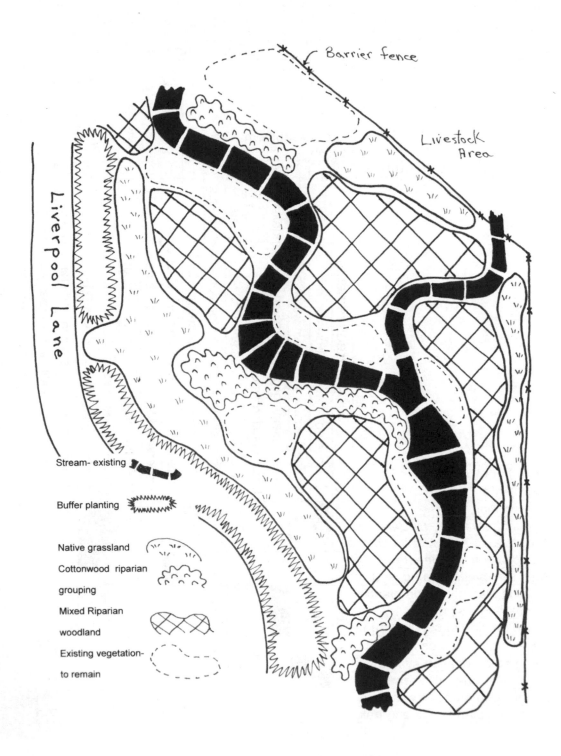

Barrier fence

Livestock Area

Liverpool Lane

Stream- existing

Buffer planting

Native grassland

Cottonwood riparian grouping

Mixed Riparian woodland

Existing vegetation- to remain

FIGURE 6-2. Bubble diagram of a restoration site.

Restoration Plan Elements	Goals and Objectives: Create southern willow scrub							Protect park and historic features			Protect native vegetation			Improve ecological functions		
Objective number	1	2	3	4	5	6	7	1	2	3	1	2	3	1	2	3
Project Requirements	Remove non-native species	Create 3 acres of southern willow scrub habitat	Protect plantings from herbivory and vandalism	Begin planting on October 15	Establish self-sustaining habitat within 3 years	Remove all evidence of restoration within 3 years	Remove top 6-8 inches of soil and leaf litter	Cause no disturbance to historic structures	Preserve 5 historic Phoenix palms near Spring H.	Minimize impacts to park trails and users	Identify all native vegetation to remain	Identify construction access to avoid damage	Identify construction staging and storage areas	Improve water quality in the creek within project	Increase diversity by returning cottonwoods	Establish native understory suitable for birds
Clearing Plan	■						■				■	■			■	■
Planting plans: so. willow scrub	■						■				■	■				
Sensitive areas noted on plan											■	■				
Access control delineated								■				■				
Palm removal at Spring House								■								
Fencing placed around plantings			■		■											
Informational signing			■													
Fence on side of service road			■													
Improve stream crossing														■		
Automated irrigation system		■			■											
Individual non-native removed	■															
Cottonwood along streambed														■	■	
Sycamore plantings along trail														■		

Met Objective - ■

Figure 6-3. The restoration plan evaluation matrix is used to verify design elements in relation to goals and objectives.

edge of your project boundaries there is a road (as in figure 6-2) or a playground that would either be a nuisance or provide an avenue to access your project site. Designing a feature that would control or otherwise address this issue would be considered a buffer. That feature could be vegetation planted to screen restoration areas from view or with sufficient distance to withstand the degrading influences of the adjacent activities. With the key project features identified, then determine the compatibility of plant associations between the various project features.

Test for Suitability

Using the Design Evaluation Matrix (fig. 6-3), you now test the conceptual design for compliance by reviewing and evaluating each of the project requirements. The matrix identifies key project elements along with key project requirements. Using this tool allows you to evaluate the conceptual design for suitability in meeting the project requirements. For example, during the initial diagramming process, a wetland area is placed along an existing streambed. Upon evaluation, you note that an existing park trail runs adjacent to the proposed wetland. When studying the project requirements, you note that the wetland is intended as habitat for a sensitive bird species and that the wetland area should be separate from human intrusions. Therefore, the proposed location for the wetland is assessed to be incompatible with the project requirement that the wetland be separate from human uses. The next step in the concept design process requires that you refine the design and either relocate the wetland area, close or relocate the footpath, or consider including a buffer to separate these incompatible features.

Refine the Concept

The final step in bubble diagramming is to refine the diagram by eliminating any incompatibilities and making the necessary adjustments to fit the proposed project features within the existing site. Here, trade-offs may be necessary as you attempt to satisfy all of the project requirements while maintaining the project within the project goals. After this step is complete, you are prepared to move into the design development phase.

Scope, Schedule, and Cost Constraints

The scope, schedule, and cost factors interact to form a dynamic design relationship. The *scope* of the project is the definition of what the project is all about: it sets the boundaries as to what the project is and what the project is not (e.g., acreage, elevations, vegetation communities). The *schedule* represents the time that it will take from project inception through completion to deliver the defined project. The *cost* represents the total cost to deliver the agreed-upon project scope. Change one factor, and adjust the other two factors accordingly, either stretching or contracting.

The key factors contributing to project design decisions are either environmental or project-definition considerations. A few examples of environmentally based factors are time, naturally occurring environmental stressors, and site resource constraints, such as the presence of endangered or sensitive species or cultural resources. Factors based on project definition include the es-

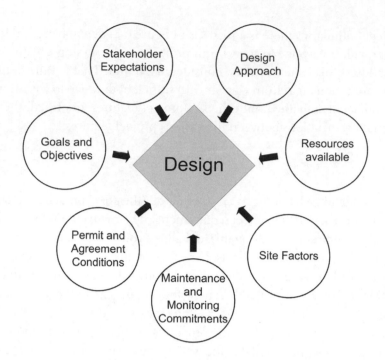

FIGURE 6-4. Factors influencing the design of a restoration project.

tablished goals and objectives for the site, the budget, evaluation criteria as established by various stakeholders, and requirements established by permitting agencies or granting organizations (fig. 6-4). These factors have a relationship with one another and, depending on the results of your site analysis and other data, will require a response in the design of the project.

Design Elements

The following list of design elements is presented as steps. Each step involves more specific issues and builds on the previous step. As mentioned earlier, this is an iterative process and a response to one factor may require an adjustment or reconsideration of a prior decision. Much of the information you will need for these elements should be evident from the reference sites or can be incorporated in the reference model prepared from reference site information. The seven elements of restorative design are as follows:

- Water body/hydrologic modification
- Land modification
- Vegetation community
- Species palette
- Vegetation patterns
- Species quantities
- Habitat features

Water Body/Hydrologic Modification

Many stream restoration projects involve the realignment or reconfiguration of a stream channel to mitigate the impacts of prior channel alterations. Sometimes there is a need to raise the bed of an existing degraded channel using various channel improvement structures to restore the hydrologic connectivity of a stream with its floodplain such that overbank flooding will occur at appropriate frequencies. The same concept would apply to lake restoration. Other times, there is only the need to return water to restore the project site. In these cases, control structures that regulate the inflow and outflow of water are sometimes needed to compensate for altered site conditions.

In coastal areas, the most common action is providing for the return of tidal flows. Once breeching a levee, the distribution of the water over the newly flooded land needs to be regulated, sometimes by channels or control devices that regulate elevation and duration of tidal inundation because in many cases the land has subsided during the time period of being isolated from the tidal waters. In these cases, it is important to have a hydrologist or surveyor determine the relationship between land surface elevations and tidal elevations at your specific location. Some riparian restoration projects involve the lowering of portions of a site adjacent to a stream channel or on a floodplain to achieve a desired frequency of inundation by floodwaters or a desired elevation above the shallow groundwater table. Vernal pool restoration projects sometimes involve excavating new pools and then lining the pools with appropriate soil materials.

Landform Modification

Some restoration projects require significant reconfiguration of land contours prior to revegetation. Major site modifications are commonly associated with lands used for material extraction (i.e., mining sites). Several decisions are required when working with abandoned material sites or other similarly degraded lands. The issues of slope, aspect, and shape of the slopes all need to be addressed when considering the vegetation to be introduced. Abandoned farmlands typically range from very flat to rolling hills and lack sufficient microelevations that create microhabitats important in providing for species diversity. The vernal pool ecosystem requires a very small scale microrelief of shallow depressions and mounds in between with very shallow slopes.

In many cases, the land modifications required on a site may not involve the actual restoration work planned but are required because the existing landform is a source of constant degradation. Over the past half century, numerous techniques have been developed with the goal of stabilizing slopes, affecting stream courses through natural means, enhancing in-stream fisheries habitat, and providing erosion and sediment control. The use of these techniques requires following a series of established procedures while using specific material to accomplish the desired goal. Topics in this category include assessment techniques, biotechnical solutions such as soil bioengineering and biotechnical slope stabilizations, and watershed stewardship. Many of these techniques have been standardized to permit the development of manuals to increase repeatable results within the parameters of the technique requirements. While standardized in how these techniques are created, the interpretation of the conditions and the selection of the appropriate technique are not standardized to the point where individuals not properly trained should be cautioned about the potential of not meeting expectations.

Vegetation Community

Projects that involve the removal or creation of land formations with different aspects and slopes affect the vegetation communities that will be restored. Commonly, over the entire range of the vegetation community there will be different species assemblages. These assemblages may be the result of the combined influences of topography, altitude, moisture availability, and soil types. These differences need to be considered and evaluated against the new landform being created or restored. As discussed earlier, most common modifications are concerned with various types of waterways and drainages. Such changes will permit the return of a vegetation community that can persist where one was not possible previously.

Species Palette

Just how many species need to be introduced to a new restoration site? Understanding the species composition of the vegetation community is an important first step in responding to this challenging issue. Next is establishing whether there is a need for specific species to serve an identified function or objective. Some species are consistently found with other species. This can be an important aspect in the overall performance of a site. Are there any sensitive, threatened, or rare species for which you will need to provide habitat or forage or which you will need to introduce directly?

The exact number of species is arguable for any site. In many cases, a species list with percentage composition will permit you to make an informed decision. Typically, about 20 percent of the species will comprise about 80 percent of the vegetation. This is not to diminish the importance of the remaining species, because restoring biodiversity should be a fundamental objective of all restoration projects. It is often in the "20 percent" that species of special concern can be found. Some of these species may serve as a food source or an egg-laying habitat for significant pollinators. Every effort to accommodate these species should be made.

Typically, the collection of topsoil with the seed bank can be introduced. The goal is to provide the presence of as many species as possible on the site. The difficult decision is how many more species need to be introduced on the site. Cost, time, and availability are major factors that will influence your ultimate decision. Prior to this, however, you should be aware of the species that are immediately adjacent to your site and the probability of those species naturally invading and becoming established on your site. Not all species can be easily propagated, nor are they available at native plant nurseries. Do you have the time to wait for seeds and to propagate them? Do you have the resources to tend to plant containers and maintain them until planting on-site? Have you planned sufficiently ahead of time to allow for a nursery to do this work for you? Can the species be added later and not delay the initiation of the project?

Vegetation Pattern

A complex aspect of design, vegetation pattern is commonly a process of self-sorting by the species selected. Common factors that should be considered initially include vegetation zonation, mosaic/cluster, edge, slope and aspect (i.e., orientation), soil affinity, and moisture requirements or tolerance. Some general factors exist to help guide some of the decisions, but not entirely. The collection of reference site data should reflect the distribution and pattern on the land of the key species. Vegetation data should provide guidance on how the individual plants are arranged on the land, information that can be applied to the design of a restoration project. The planting layout must consider slope and solar aspect to ensure that sun-tolerant species are placed in appropriately sunny and warm areas and that species which require cooler and moister conditions are placed on slopes less exposed to the sun. Likewise, species that are less tolerant to extreme sun and heat should not be placed on west- and south-facing slopes.

ZONATION

Zonation is a common vegetation pattern of bands or other shapes that are frequently influenced by elevation, moisture, soils, or the effects of plant competition and that may occur either over a large geographic area or in small areas. Zonation is easily observed in wetland situations from the water surface to the tops of riverbanks and higher elevations. This pattern is observable in small prairie patches, vernal pools, or seasonal wetlands with an elevation change of as little as six inches. Upland species also exhibit a zonation pattern, although it is not always as easily detected as in wetlands. This pattern commonly occurs on hillsides in drainages, canyons, and ravines. The vegetation may be responding to gradients in climate, moisture, elevation, slope exposure (north versus south facing), or other environmental factors.

MOSAIC/CLUSTER

A mosaic/cluster pattern can be found in many vegetation communities. The sources for this pattern are many. Soil types, compaction of soil, availability of moisture, seed distribution, seed viability, timing of rainfall, resistance of seed to periods of drought, predation on seed, and herbivore preference in grazing are some factors contributing to species sorting in the vegetation. Urbanization has also contributed to this mosaic pattern with the introduction of exotic landscaping that contrasts dramatically with the adjacent wildlands. These urban landscapes have divided up the vegetation into small, isolated colonies of native vegetation, causing fragmentation of the resources and creating less optimum habitat for native species.

EDGE

In some vegetation communities and regions of the world, the important aspect of the site, to wildlife, is either the amount of edge to area between contrasting vegetation types (ecotone) or the extent of homogeneous vegetation. Figure 6-5 illustrates three examples of configuration in a mosaic or cluster situation. The edge in figure 6-5a is significantly higher than that in figure 6-5c even though there is the same amount of vegetation area for the two areas illustrated. This situation may be important for

FIGURE 6-5 A,B,C. Three examples of equivalent edge to area designs illustrating the importance of design to approximate specific habitat requirements of target species.

wildlife species requiring edge for some aspect of their life history. Then again, there are numerous species that occur in more homogeneous areas and avoid edges. Knowing this habitat requirement for your target species is especially important when designing your project. If your site is of insufficient size, you may want to focus on reducing edge because the distances within the various habitats on your site may be insufficient to have the desired effect on targeted species. Increased abundance and diversity of wildlife along edges frequently increases biotic interactions, such as predation, brood parasitism, and competition. Depending on your goals, these outcomes may be acceptable.

Knowing your target species or species guild will assist in the decision-making process of design (Morrison 2009). In many cases, natural history studies are available that provide sufficient details from which a design can be formulated. However, this is not always the case, such as for the work the senior author conducted on the endangered least Bell's vireo in California, for which detailed natural history studies did not exist. So, in that case, habitat studies were conducted on three rivers that comprised a little over 10 percent of the known population at that time. These data were synthesized into a design model that included plant counts, plant heights, and pattern within a quadrant. The results of these field studies were used successfully to create several projects (figs. 9-3 and 9-5) (Baird and Rieger 1989).

SLOPE ASPECT/ORIENTATION

The duration of direct sun on various slopes contributes to the occurrence of vegetation communities. Some communities occur consistently or have healthier growth on a specific slope aspect. Moisture is believed to be the primary factor controlling this distribution. Following the seeding and planting of containers, we have observed a shifting of vegetation in response to a pattern of inland sage scrub preferring east- and south-facing slopes and chaparral occurring predominantly on northern and western slopes.

SOIL AFFINITIES

In many cases, a complex of species has adapted to unusual or uncommon soil conditions. Sometimes the distribution of soil may be a few square meters, or other times it may cover hectares. Knowing if you are in a location with unusual soil and species associations adapted to those soils will be critical in the planning of where and how to obtain specimens. Serpentine plant community is found primarily on serpentine soils, and several species in this community are considered rare because they occur only on these soils. In situations such as this, it is important to determine the subspecies or variants; it is not uncommon for species to have locally adapted forms. This is very important to project success because available inventory may not grow adequately on your site.

Moisture Requirements or Tolerance

Moisture level strongly influences the sorting of species in the vegetation. As previously discussed in zonation, this moisture level is more general and manifests itself as isolates or colonies within a larger, contrasting vegetation community. Seeps, or perched water tables, often contribute to this situation. Areas with several soil types with varying abilities in water retention will also permit some species to persist in an otherwise unsuitable location. Small pockets of soil can create significant changes in the vegetation. In some cases, plantings will not survive, because the extremes of the moisture exceed the tolerance of the plant.

Given all of these factors, you still have several decisions that must be made to complete the design of a project. Studies on islands have demonstrated a relationship between larger size islands, in either land mass or vegetation area, and an increase in

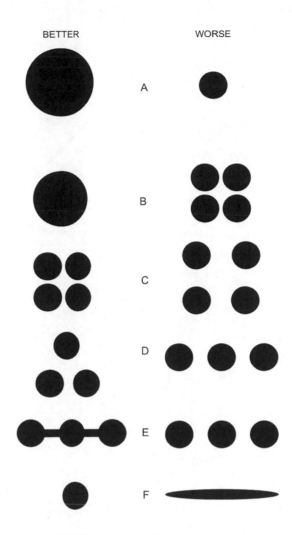

Figure 6-6. The ease and frequency of wildlife movement from one island to the next is determined by distance and size of the island. Occupation of an island is also influenced by the configuration. (Adapted from Diamond 1975 and used with permission by Elsevier Science.)

species diversity. Using the findings of these studies and his personal observations in New Guinea, Jared Diamond (1975) developed a hypothetical configuration graphic that illustrates immigration potential (fig. 6-6). The preferred geometric design typically has more area to edge, less distance, or a connection by corridor. When working on large tracts of land, it may help to approach your restoration projects with these models in mind. The size of these islands also factors into the expected species supported (Howell, Harrington, and Glass 2012).

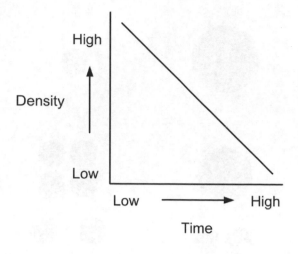

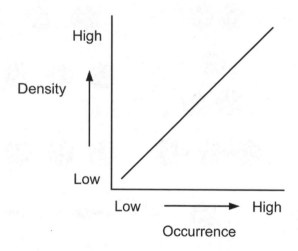

FIGURE 6-7. *(Top)* Site plant cover can be attained sooner with high planting density; however, for some species, higher density (i.e., overcrowding) will negatively affect plant growth and stature.

FIGURE 6-8. *(Bottom)* Higher planting density and the occurrence of competition with exotics. Higher-density planting is a high competition level for space and light, which frequently will control densities of exotics.

In many instances, we are involved in project sites that do not afford us the luxury of using these models entirely; however, they have some relevance even on the small scale that is typical for restoration projects (Morrison 2009).

Species Quantities

The number of individuals planted or seeds sown per area is influenced by the factors mentioned earlier: evaluation criteria, time to meet goals and objectives, and stakeholder expectations. In addition to these criteria are the specific site stressors: Is the site subject to extremes of temperature or rainfall? Does the potential exist for significant numbers of exotic species to invade the site? Does the budget provide for only a minimum of maintenance? Other factors that may also impact your project are the potential for vandalism, the unavailability of species due to rarity or uniqueness, a naturally high mortality of plants selected, and a poor seed crop the previous year.

Typical reasons for planting or sowing at higher rates are to prevent or retard the establishment of exotic species by creating a denser plant cover earlier; the site may be subjected to various herbivores, from insects to deer; maintenance is not likely to occur on a regular basis; and the site has a known pattern of experiencing extremes in one or more natural processes.

The four graphs (figs. 6-7, 6-8, 6-9, and 6-10) illustrate the relationship between planting density and some of the most frequent factors involved in restoration projects. However, these graphs represent only one factor in an idealized situation. In reality, you will be required to address several factors and to make compromises. For instance, by providing a high level of maintenance in figure 6-6d, you should be able to control the unwanted establishment of exotics. So, if these two factors are of concern for your project, then by providing maintenance you simultaneously address the exotics issue, but only if the maintenance budget includes weeding. Given this situation, you can design a project with less dense plantings. However, if your permits or stakeholders are expecting a certain level of performance, you will have to incorporate this into your design thinking. In the area of environmental stressors, your site analysis may have addressed a number of these and the overall construction plan may be removing some or all of them. The remaining stressors may have a bearing on your planting density decision. From this simple set of graphs you can visualize the relationship that the various factors have and how they will influence your decision on the final design.

These examples are designed to illustrate the relationship of one factor to another. In the final analysis, the ultimate aim is to have the planting attain a level of self-sufficiency and then further develop and mature into the prescribed type of vegetation community. While the drawings of the plans may show discrete circles, polygons, and similar symbols, the intent is not that these species should only remain or even continue in these locations. We are not interpreting the plans as if they were landscape architecture drawings that are meant to be site specific with little to no variation. Our ultimate purpose is to introduce the species onto the site and permit the species to survive wherever they find suitable microhabitat. We start the process by making determinations where we

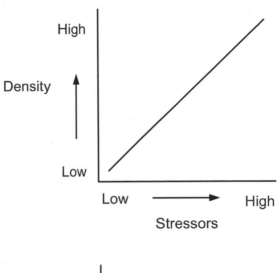

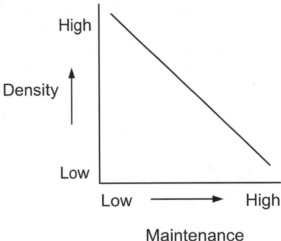

FIGURE 6-9. *(Top)* In areas with significant stressor influences (temperature, moisture, wind, insect infestation), a higher than normal planting density will increase the probability of a sufficient number of plants surviving.

FIGURE 6-10. *(Bottom)* Sites with a high intensity of maintenance will permit lower planting density because stressors will be controlled, increasing the likelihood of higher plant survival.

feel the various species will be best suited, but in the end the species themselves will tell us where the suitable sites are located.

Habitat Features

Thus far in this chapter, we have addressed vegetation elements and aspects. Habitat features are focused on specific physical shapes and patterns and their attributes (fig. 6-11). Habitat features can be formed by parts of plant material or by the physical arrangements of rock and soil. These features provide a spatial context that attracts individuals of a species. A goal of establishing habitat for a species requires that all the elements appropriate for the habitat be present on-site for the species to become a resident (Morrison, Scott, and Tennant 1994). Or, in the case of providing wintering

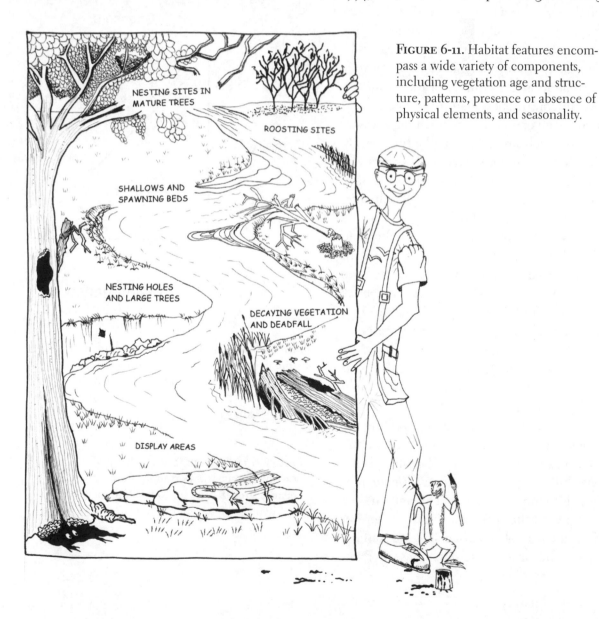

FIGURE 6-11. Habitat features encompass a wide variety of components, including vegetation age and structure, patterns, presence or absence of physical elements, and seasonality.

FIGURE 6-12. Nest boxes provide immediate nesting sites for many species ranging from small songbirds to ducks. Patuxent Research Refuge, Patuxent, Maryland. (Photo by Mary F. Platter-Rieger.)

grounds or some other function, those elements that satisfy the requirements of the species must be present. In many cases, a feature is something that is developed over time. If you have the time, then you do not need to actively provide it from the initial establishment of the site. However, this is not typically the case, and you must be resourceful in providing those habitat features.

A good example is cavity nesters, both birds and mammals. Unless there are old mature trees, rocky outcroppings, or a steep embankment, it is unlikely that a newly constructed site will have these specific features. The latter two examples can be constructed on-site with fairly routine methods. However, a tree cavity requires a large tree. Does the tree have to be alive or dead? Does it have to be a tree? Will the target species use a substitute structure, such as a bird nesting box (fig. 6-12)? The latter, of course, would require some maintenance, but as an interim solution, this approach has proven successful for several bird species in the eastern United States. In the western United States, burrowing owls have benefited from using nesting sites constructed from PVC pipe and utility boxes. Nest boxes for barn owls have been used with great success along the Sacramento River in California. These should be considered interim measures until the site can mature. In the case of the owl, they typically take over holes initially dug by ground squirrels. If no squirrels are on-site, this interim solution will provide time for the site to develop and a squirrel population to become established to create future nesting opportunities for the burrowing owl.

Providing these types of interim solutions may be necessary to ensure the continued presence of the targeted species during restoration activities. Although artificial in many respects, this may be the best method to allow the species to remain in the project area. If it is likely that the target species may not reoccupy the site following restoration activity, then these artificial measures will permit the presence of the species until suitable natural habitat has been created or provided on the project site.

FIGURE 6-13. As an interim solution for the loss of perching sites along the river, salvaged branches and trunks were stuck into the ground. Hawks, herons, and egrets were observed using the perches year-round. San Diego, California. (Photo by John Rieger.)

Habitat features can address several functions in the life history of a species. The most common is to provide a structure or vegetation arrangement conducive to reproductive displays and mate attraction behaviors, territorial display and singing, as well as nesting, denning, or bedding sites.

Vegetation and physical features also provide shelter for prey. Foraging sites and vantage points for predatory animals may also be required, but given the more mature state of these features, it may be necessary to put in interim structures that can serve the same purpose (fig. 6-13). Do water features have a role in the life cycle of the target species or the assemblage of species anticipated to use the site? Plants may be the required food of a species. Knowing the needs of the target species is important if the entire habitat is to be restored and a self-sustaining population is to be established on-site.

Another important feature is providing refuges for specific animals, whether these are narrow rock crevices, tall trees, or high ground during flooding. Is there a need to provide an escape area during a catastrophic event or simply in the normal order of the day? Do prey animals have a place to elude capture?

By evaluating each of these issues and giving attention to the specific needs as outlined in the goals and objectives statements, you will be able to develop a sound design with a minimum of confusion and wasted effort. Depending on its size, a project may require only the simplest of design considerations. However, for even small projects, you will find that your needs will involve more than just a few of the options that have been discussed. Again, as with most decisions made in the process of developing a project, your foundation is the goals and objectives that you developed prior to this phase of the project. Another major filter process is the budget and time. Can you afford this type of design or this approach to meeting the goals? There are several variables that you control, some with more substantial effect than others. It will be your decision on how to proceed and what variables to modify.

One example of this situation is the decision made to build a fence around a significant part of the restoration area within Los Peñasquitos Canyon Preserve, in San Diego, California (chapter 14). The fence will keep hikers and mountain bikers out of the newly planted area, thus reducing maintenance cost, including the replacement of plants. For this project, fencing was the most cost-effective approach. In addition, and more importantly, constructing the fence met stakeholder expectations by keeping loss of time to a minimum and site growth to a maximum.

Chapter 7

Water and Soil

Restoration projects involving plant material will always require knowledge about the requirements for water and the suitability of the soil as a growing medium for the plant materials. Whether you are concerned with wetlands, forests, scrublands, or even deserts, you must know the historical aspects of water and soil conditions. Surprisingly, a number of wetland restoration/creation projects have been implemented with scant attention given to landform and elevation. As a result, the surface elevations of these project sites have either been too low or too high with respect to tidal influence or streamflows. The end result is a project that fails to perform as intended. Soils are complex assemblages of organisms, mineral sediments, water, and atmosphere that together create a growing medium. Typically, intensive farming depletes nutrients or introduces excessive amounts of nutrients that are not needed by native plants or that attract weeds and exotic species. Water and soil are two basic mediums that cannot be readily assessed simply by looking. To verify your supposition, a few laboratory tests are required to ensure the chemical constituents are within acceptable ranges for your intended purpose.

New plantings and sowings will typically require water before becoming fully established in arid regions or those with dry seasons. Requirements of the plant species used will determine the duration and timing for water application, especially in areas with Mediterranean or desert environments. Planning for water cannot be overemphasized in those regions. If you are able to install your plantings during the appropriate rainy season, this strategy may provide the best use of resources and labor. However, if you are not able to accomplish the needed work in the short time prior to or during your rainfall season, then evaluate alternate strategies. Obviously, if your project is involved with wetland ecosystems, then addressing water supply issues in detail will be required to ensure project permanence. However, in dry landscapes, water may need to be applied until the natural rainfall cycle begins or is sufficient to support your plantings. The decision to provide supplemental water to your site will guide subsequent design and site preparation decisions; the absence of supplemental water may affect the development rate.

Water

Site suitability and the time of year during which the plant materials are installed will determine the need for a supplemental water supply. If you are modifying existing topography adjacent to a water source, and you have determined the water regime so that the site will be exposed to sufficient water shortly after planting or seeding, then supplemental water may not be necessary. However, if the site is elevated and intended for only occasional flooding, as in the case of benches or terraces adjacent to streams and rivers, then providing water to allow for root growth will need to be addressed. If you have control over the planting or seeding time, and the rainfall or flooding is fairly predictable, you may have an acceptable risk level for not providing supplemental water. Obviously, wetland systems operate under a different hydrological regime than uplands. You may need to calculate the water budget for your project to ensure that sufficient water will be available to promote plant growth (Pierce 1993).

Understanding how the hydrological factors are at work on your project site is necessary to ensure successful plant establishment. Water can occur on your site naturally through natural watercourses, ponds, or seeps. Water in the subsurface, commonly called groundwater, tends to fluctuate with the season. Water from rainfall changes watercourses or changes the level of water bodies, such as ponds or lakes. Knowing how water behaves on your site will influence several decisions you must make. How will you go about restoring your site? When will you attempt the restoration effort? What material or equipment or agreements will you require?

For projects designed to use only naturally occurring water, the timing should coincide with the time of year when water is available and plant growth will occur. Dormant plants allow for a planting program without the need to have water available until the breaking of dormancy. The instal-

FIGURE 7-1. Imprinter device used for creating catchments for seed and water to promote plant establishment on coastal slopes. Different designs of the imprinter are used extensively in arid land situations. (Photo by Mary F. Platter-Rieger.)

Box 7-1. Lesson Learned: Be Sure Substitute Equipment Is Truly Equivalent

Land imprinting is a technique of using a device that leaves small catchments of a specific configuration designed to concentrate water, seed, and micronutrients. Used primarily on lands that are flat to rolling hills with limited water, these devices can be used on slopes up to 2:1. The senior author, initially not familiar with the intricacies of the land imprinting process, was persuaded by engineers into allowing the use of a swamp cat treaded bulldozer instead of an actual imprinter. Although the individual imprints approximated the imprinting device, there was no system of concentrating water into small reservoirs at the ends of the track troughs. Instead, the water built up and eroded the small wall of the trough, and numerous erosion rills were formed.

In another example, the senior author was challenged with the moving of mature trees. Superiors determined it was not necessary to use a standard tree spade so the contractor used a standard backhoe. The failure was almost immediate due to an inadequate rootball and a major disruption of the rootball by the backhoe bucket.

Lesson learned: The application of equipment designed for one purpose may not translate completely to the needs of your restoration project. Substitutions to planned work are not uncommon. When introducing a new technique or piece of equipment, be completely familiar with its details and how it separates itself from similar equipment or techniques. When a change is proposed, discuss the proposal with your team for careful evaluation before approving the change.

lation of many projects can be timed to take advantage of a predictable weather pattern and avoid the need to provide an elaborate system to deliver water to the site. Evaluate the size of the project and the duration of the "wet" season. In situations with unpredictable or unstable weather patterns, the use of seeds may be the best choice and plant mortality should be expected at higher levels than with supplemental watering. This is acceptable as long as you have planned for this approach.

Knowing the growth characteristics of the species will greatly assist in your decisions. Root growth behavior is an extremely important trait when relying on natural systems. Your design and timing should account for soil moisture in the root zone during the growing season. Arid lands rely on storm events to recharge the soil column. To enhance germination results, an imprinting device is used to create numerous surface depressions. These depressions concentrate water, keep seeds on-site, and collect debris and micronutrients. Erosion within the depressions buries the seed and reduces loss by wind and seed-eating animals (Bainbridge 2007). Survival of germinating plants within these depressions is higher than in adjacent areas. The imprinter is effective on slopes to reduce erosion, contain seed on the slope, and promote plant growth. Imprinters are designed for the function they will provide, so a wide range of catchments and water collection systems can be devised. Figure 7.1 shows a small imprinter designed for tight-access slopes. Several attempts have been made to use standard caterpillar tread tracks to substitute for an imprinter, but with unsuccessful results (box 7-1). In addition to creating small water collecting depressions, other options are available to concentrate

Table 7-1. Design Features Planning Analysis and Obligations for Supplemental Water Systems

Topic System/Technique	Requirements to Use	How Determined	Planning Obligation	Operations Required
Existing Water: Groundwater	Present and reachable by plant root systems. Soil not compacted. Depth is fairly consistent or has predictable changes from natural or regulated human use in watershed.	Dig wells; drill holes for water depth measuring devices or observations. Other sources from adjacent landowners, existing wells, and agricultural agencies.	Run wells or data collection in preliminary stages of design.	None required.
Existing Water: Watercourse	Constant or predictable flow.	Collect flow data from gauging stations and water departments.	Use collection data from existing sources or generate own data set.	Ensure delivery of water in system by others, if required.
Existing Water: Lake	More or less constant elevation.	Lake owner elevation data, physical evidence of vegetation denudation in backwater, presence of mudflats.	Confirmation of lake status and potential uses of lake water that may affect elevation.	None required.
Existing Water: Seasonal Rainfall	Predictable seasonal rainfall within normal intensity anticipated.	Weather data from nearby.	If data unavailable, establish rainfall monitoring station.	Plant during appropriate time in rainfall cycle.
Supplemental Water: Canals/Piping	Water elevation sufficient for gravity flow.	Topographic maps for general data; for short distances, use a surveyor's level.	Determine source of water; purchase or secure rights.	Maintain and monitor canal and water control devices.
Supplemental Water: Cistern Systems	Sufficient distribution systems from cistern. Elevation for gravity feed.	Same as above; also determines delivery system within site. Small ditches, piping, or general flood irrigation systems.	Similar to above; in addition, plan on a water delivery system to cistern. Catchment structure for rainfall or water truck, hand bucket brigade.	Control flow from cistern. Maintain pipe or ditches from cistern.
Supplemental Water: Pressurized System Piping/ drip/bubbler	Access to established water system.	Pipeline or water tower nearby for access. No surface water available. Rainfall generally unpredictable.	Generally requires an agreement from a water utility company. If flood irrigating, energy-dissipating devices may be required.	Monitor and maintain system. Devices require cleaning regularly.

or bring water to a site. Several systems have been developed for use in arid and semiarid regions of the world (table 7-1) and have been discussed more thoroughly in Bainbridge (2007).

Container plants offer a different challenge because they have been supported in the nursery and will experience stress during the planting process. Placing container plants in the ground immediately puts the roots in a stressed condition. Will the soil contain sufficient moisture to support plants with very shallow roots? Will there be sufficient time for the roots to penetrate the soil down to more reliable sources of water, such as a stable groundwater supply? Is the soil sufficiently loose to allow for normal root growth? If the site was excavated down (i.e., lowered in elevation), did the heavy equipment cause compaction of the upper level of the soil that can retard root growth? Whatever your decision is regarding the use of water, be sure to think through the needs of the living material you will be selecting for the project.

Developing a Water Supply

The decision to provide supplemental water to your site will guide subsequent decisions in the design and site preparation aspects. Difficulty arises when large tracts of exposed ground are susceptible to erosion. Will the proposed design alter the topography, making some portions of the site more accessible to water sources, or will a new water supply be developed? Table 7-2 illustrates how the various water sources can be manipulated to the advantage of your project. The possibilities are almost limitless given the diversity of conditions and situations in nature.

Table 7-2. Water Supply Development Approaches

Natural Systems and Processes	Element	Action to Exploit Resource
Precipitation	Rain, snow, sleet, hail	Improve permeability of soil; increase roughness of ground to catch water and confine on-site; create small basins or rows to concentrate water.
Surface Runoff	Unconfined water flow overland	Install small basins (i.e., imprinting), dikes and berms, ditches, gullies, ponds, below-ground cisterns/tanks.
Groundwater	Saturated subsoil	Excavate original ground to bring closer to table level.
Existing Water Body	Stream, creek, river	Create diversion canals; install piping using gravity flow; widen channel and bank; flatten banks to increase area.
	Pond, lake, sea, ocean	Excavate out shoreline; fill in shallows to raise bottom; create canal or channel.
	Seep, springs	Channel water; flatten slope to expand area.

FIGURE 7-2. Constructed vernal pools situated using data collected on watershed to pool area ratios. Rainfall is the only means of filling these seasonal pools. Seeds, collected from adjacent natural pools, were hand broadcast. San Diego, California. (Photo by John Rieger.)

Determining the Type of Water Delivery and Storage System

There are many methods of delivering water to your site (figs. 7-2 and 7-3). Depending on the source and the orientation of your site, some methods may be very simple (box 7-2). If the water supply is below your site, then getting the water up to your site may require pumping or some other type of system.

Some options have only one method of delivery, whereas others that rely on the natural weather patterns will require the development of storage systems so water can be made available to the site outside of the normal precipitation season.

Maintenance Requirements

The initial system selected will significantly affect the activities required through the life of the water supply system. For short-lived systems, this may not be a huge impact; on large projects, however, regardless of duration, it may strongly affect your resource budget. Some of the most common maintenance issues include the following:

- Labor: repair of facility, grading, digging, valve repair, engines on pumps
- Fuel: operate pumps, water trucks
- Water flow: use hand or heavy equipment to repair dikes and levees
- Pest control: manual and chemical vegetative control of waterway, trapping, pesticide treatments, exclosures for animal control
- Operations of water system: regulating delivery, gate control, siphons, and so forth

FIGURE 7-3. A storage and gravity distribution system provides water in the Colorado desert. A cattle trough has been placed under a bridge abutment to provide gravitational head. The water travels through PVC pipes to buried unglazed clay pots. (Photo by David A. Bainbridge.)

Box 7-2. Creative Solution: Low-Tech Irrigation System

A contractor was responsible for the design, installation, and maintenance of plantings adjacent to a small lake. To save money, the contractor avoided installing an elaborate irrigation system. However, the plantings required watering several times through the summer of the first year. The contractor used a portable pump to pump water from the lake to the plants. The contractor found that it took a very long time to water each plant using a hose to distribute water to each planting basin. To reduce the time required to water each plant (refilling the water basin more than once), the contractor installed a five-gallon plastic bucket next to each plant with a small hole at the bottom side of each bucket. The maintenance workers were able to rapidly fill each bucket and minimize the time required to water each plant.

Costs Associated with Water Delivery and Storage Systems

Having made the design decision on how water will be handled on your project, it is now possible to analyze the various elements of the system related to resource costs. This includes not only actual monetary expenditures but also labor, materials, and potentially governmental regulatory costs.

Four of the most common costs associated with water delivery systems are listed below:

- Developing the water source
  - Grading equipment
  - Water control gates and barriers
  - Earth-moving equipment if beyond scope of the grading equipment
  - Pumping supplies, including pumps, pipe, and valves
  - Appropriate governmental permits and clearances
  - Labor
  - Surveying of site to establish elevations
- Delivering the water
  - Grading for ditch system, canals, pond, and dikes
  - Labor
  - Tanks or other storage devices (if not pond)
  - Pumps
  - Electrical systems to operate pumps (if not electrically operated)
- Operating the system
  - Labor
  - Energy costs, electrical or fuel for pumps (if not electrical)
  - Water charge cost for commercial providers of potable, reclaimed, or agricultural
- Maintaining the system
  - Labor
  - Replacement of equipment and supplies
  - Equipment to recontour pond, canals, or ditches

Depending on the constraints and objectives of your project, many items may not occur in your design. Simpler, in most cases, is generally better.

Natural Systems and Restoration Projects

Upland areas have different characteristics depending on their location in the overall topography and watershed. Knowing the rainfall pattern and amount will be necessary to ensure that a site will be permanent. Sites on flatter slopes and adjacent to watercourses but not directly flooded by these courses may receive water by the rise and fall pattern of groundwater as it is influenced by the watercourse. In extremely dry areas, the groundwater may originate primarily from the rainfall and flooding events. The duration and intensity of these storms are critical to providing a reservoir of water during the growing season so that plants can send root systems sufficiently deep to stay alive. As water becomes unavailable at the upper levels of the soil, roots are able to move down the soil column following the receding moisture. This trait improves the survival probability of most desert plants beyond the rainfall season. Some arid land plants, such as mesquite and juniper, are able to drop roots in excess of one hundred feet below the surface.

Duration and frequency of irrigation are the most important components of an irrigation strategy. Soil type and temperature are major environmental factors that determine the frequency and duration of irrigation. Watering too long at a time will provide a saturated soil condition near the surface and not force the root systems to grow downward. Shallow roots are typical for some species, such as willows, which are typically found near flowing streams with groundwater several inches below the surface. The preferred irrigation schedule is to create a pulsing regime whereby the water column is not totally saturated but is allowed to become drier near the surface as the water moves downward from evaporation, transpiration, and infiltration. This method of irrigation facilitates root growth downward toward more permanent water sources.

Frequency of watering is another critical factor for successful growth. The variability of the site and the many environmental factors on-site require that you examine the plants and let the signs exhibited by them determine your watering frequency. In many arid regions, water, seeds, scarification, and absence of predation occur simultaneously, permitting the establishment of plant species. Restoration of an arid area outside of this simultaneous occurrence will require consideration of supplemental watering and how it interacts with the existing porous soil. In many cases, a single major storm event is enough to load the soil column with sufficient water to supply the young plants and the root systems that rapidly grow down to significant depths.

The movement of water is much faster in soil with high sand content. In some soils, the water may actually pass through the root zone so quickly that it is not available long enough for plants to benefit.

Water Chemistry

Water chemistry is often not considered in the early stages of project planning. The existence of a lake or stream on-site is no assurance of suitable water. Our example in chapter 14 describes the challenge of using pond water following a five-year "drought" causing a buildup of salts in the adjacent pond water. Soil type and duration of watering are factors affecting whether it is appropriate to use water with unusually high concentrations of salts.

Commonly, situations exist that preclude one from closely following the principles of ecological restoration. Modified terrain and the presence of human habitation sometimes dictate making adjustments and adapting to the current conditions. Such is the case in Jamaica Bay, New York, where a significant program of projects has been under way to clean the water of the bay. The municipal wastewater and stormwater systems are contributing to the decline in water quality and negatively affecting benthic and other animals throughout the bay (box 7-3). The magnitude and severity of the challenge demand some rather novel engineering solutions as well as adapting biological resources to contribute to restoring the water quality of Jamaica Bay. In addition, several projects have been implemented to retard shoreline erosion and establish marshlands to maintain the shorelines at several locations within Jamaica Bay.

Box 7-3. Restoration Project Highlight: Ecosystem Restoration May Require a Multifaceted Program Covering a Wide Range of Biological and Engineering Techniques and Activities

Location: Jamaica Bay Watershed Protection Plan, New York, New York, United States

Jamaica Bay is an eighteen-thousand-acre body of water adjacent to the communities of Queens and south Brooklyn of New York City. During the past one hundred years, the bay has undergone significant changes that have disrupted its natural ecosystems. The significant population increase of New York City has contributed to increased levels of pollutants and sewage discharged into the bay. New York City, like numerous other cities on the US east coast, has a combined sewer and stormwater system that frequently gets overwhelmed during intense storm events, causing sewage to enter Jamaica Bay at a number of outlets. Another impact to the bay has been from extensive dredging during the past century, resulting in a bay bottom that no longer has the topography and substrate that can support extensive benthic organisms. As an outcome of this dredging, the water currents and hydroperiod of the bay have been significantly altered, causing erosive activity on numerous shores of the bay and its islands.

In 2005, the New York City Department of Environmental Protection initiated a multifaceted approach to solving the Jamaica Bay pollution and environmental degradation problem. After realizing that engineering solutions alone (not discussed here) were not going to solve the problem, the study began exploring innovative environmental engineering and ecological-based actions that could help reverse the pollution problem.

Of immediate concern within the bay has been eutrophication. One of the most prevalent problems is sea lettuce (*Ulva lactuca*) that dislodges from the bottom and creates large floating mats over a significant area. These mats eventually sink to the bottom to smother and kill the benthic organisms on the bay floor. To control this problem, the city successfully tested their trash skimmers to collect these mats.

Sewerage is a problem that requires several approaches because it ends up in the bay and causes eutrophication. Used in many locations outside of New York, the algal turf scrubber system is essentially a chute with an algal-encrusted bottom that cleans the water as it passes over the algae. Typically, the water traverses a channel 100 to 150 feet long, depending on the amount of algae growing in the chute. The length is typically determined by the location of the device and the native algae. Trials were designed to confirm the effectiveness of the system and to fine-tune the physical structure for maximum effect. The cleaned water is then discharged back into the bay.

A second method currently being evaluated is the use of a ribbed mussel bed. The mussel beds function to filter the water coming into the bay. As a natural system, the mussel beds can expand naturally and produce more individuals serving to filter the water. The mussel beds can be located in different areas within the bay itself—at stream outlets and other sources of pollution. Beds are located in areas not frequented by humans; the mussel is not used as a food source by humans.

To enhance and restore various biological resources in the bay, efforts have been made to establish oyster beds and eelgrass beds, and to stabilize or increase existing islands through shoreline protection. Oyster beds require stable substrate to attach. Concrete balls were inoculated with spat (young oysters) and grown off-site in favorable conditions prior to being placed in the bay. After two years, growth is continuing on the balls and this technique is considered successful. The filtering action by oysters will contribute to cleaning and polishing the water and will provide improved habitat so that the eelgrass (*Zostera marina*) beds can become established. Eelgrass was planted in a few locations to determine whether it would thrive. The results were not conclusive; most plants were alive but not thriving. The conclusion is that the water is not clear enough yet for the eelgrass to establish strong beds at this time. The idea is that if a colony can get started, it would then improve the habitat so it could be self-sustaining.

The natural islands occurring in the bay have been eroding at high rates since the dredging activity and sea level rise. The steep slope and depths of the bay bottom are such that using traditional methods of rock protection were not considered viable. To protect the shoreline, a floating device was needed, because the slope of the bottom is too steep. The solution was a wave attenuator system; these items are floating platforms anchored by chain to the bay bottom and located offshore of the islands. This positioning serves to break up the energy of the waves and currents. This interruption of the waves also causes the sediment to drop out and eventually raise the elevation of the bottom and promote a change in topography around the islands.

Stormwater management is a major focus for watershed actions. Several actions have been identified that could be used in the watershed with varying degrees of effectiveness and practicality. Because no one system will function the same in all areas and the circumstances may change from street to street, a method was needed to aid the planners, designers, and decision makers in the best management practices. That method or tool was the Ecological Atlas (EcoAtlas) developed by their consultant. It is a compilation of data throughout the watershed with associated photographs using a GIS-type format so that the information can be used to maximum extent.

The stormwater management features will include enhanced tree pits, street-side infiltration swales, bioretention areas, wet meadow, porous asphalt, and FilterPave in addition to blue and green roof features. Pilot studies for several of these features are under way, with data being collected for comparative purposes and evaluation.

In addition to these various activities and pilot programs, the city has been supporting through direct and matching funds the active restoration of degraded marshes around the bay. Several projects have been initiated in the past and are included in various management plan programs for the future. The work is done by a wide range of organizations and volunteers as well as involving the communities and related agencies.

This broad program of projects and actions is designed to clean and redirect the water, and delay its path to the bay to prevent overwhelming the sewage and stormwater system. Some of the activity is designed to help start reestablishing specific benthic organisms that filter the water to create suitable areas for benthic populations to become established. While it is not feasible to return the bay bottom to its original elevation, maintaining and enhancing the remaining islands will afford significant habitats for future colonization and visitation by mobile species.

Soil Considerations: "Don't Treat Your Soil Like Dirt!"

The suitability of soil cannot always be determined from visual site inspection. Problems with soil may have their origins decades in the past. Past land uses can dramatically alter soil chemistry and structure. More often than not, soil testing will be required to ensure that the appropriate soil chemistry is present. It is worth the time and money to conduct some chemical and physical testing. Disasters can be avoided by a few simple tests.

Soil is the unconsolidated mineral and organic material on the immediate surface of the earth that serves as a natural medium for the growth of land plants. Soil is a "living system" of numerous organisms that contribute to its formation and maintenance. Organisms residing in the soil substrate are diverse and include algae, fungi, protozoa, nematodes, gastropods, arthropods, earthworms, and small mammals such as voles, shrews, and gophers.

Soil is composed of solids and pore space. In the soil matrix, the pore space is occupied by air and water. Data addressing appropriate water, air, and soil composition for native vegetation are very scarce because research in these areas for nonagricultural lands has not been a focus until recent years. In most texts, the figures of ideal percentage of water, air, and soil are derived from optimum agricultural land and therefore, most likely, will provide only a partial awareness of requirements for your project.

Among the host of different organisms found in soil, mycorrhizae have drawn a great deal of attention. Several companies advertise inoculant for sale, and others sell plants inoculated with spores. The advantages of mycorrhizae are significant for most plants; however, not all plants respond in the same way. Pioneering plants are able to do well without mycorrhizae (Allen 1991). Spores are carried in the wind and by animals, especially insects. Wind has been found to be the most significant dispersal agent for spores, which are frequently trapped by existing plants. Disturbed soil will have spores, but the network of fungi hyphae may no longer function efficiently. Recently graded or denuded land will be devoid of mycorrhizae. Sources of inoculums can be found in undisturbed native soils with native vegetation. This is the best source of obtaining mycorrhizae for your project because it will most likely contain the species indigenous to your region. Unless your project is several hundred acres or is not adjacent to native vegetation, you likely will not need to be concerned about providing mycorrhizae through deliberate actions of your own. The vegetation community, the condition of the adjacent populations, and the substrate all factor into the ability of mycorrhizae to colonize the restoration site (Harris 2009).

The chemistry of soil is diverse and reflects exposure to local conditions. Determining variations in the concentrations of various soil constituents is not typically possible by visual inspection and thus requires a series of tests. Soils occurring in intense industrial areas that had smokestack emissions often have very low acidity and a high concentration of metals. Work in these areas requires thorough soil testing to ensure that certain elements, typically found in small quantities, have not been increased from the industrial activity. Agricultural lands, active or abandoned, may also have chemistry imbalances resulting from the farming practices and the crops grown. Physical and chemical problems are commonly encountered on candidate restoration sites. Either the primary cause or secondary effects of degradation problems with the soil should be addressed prior to developing elaborate plans.

Soil compaction is one of the most common problems on abandoned lands (Luce 1997). Soil compaction can result from trampling by livestock, repeated pedestrian foot traffic, the use of heavy equipment associated with agricultural and forestry practices, or unrestricted vehicle traffic and parking. Soil compaction is the process of increasing the density of soil by packing soil particles closer together, causing a reduction in the volume of air. Soil water acts as a lubricant increasing compaction when a load is imposed on the soil. By packing primary soil particles (sand, silt, clay) and soil aggregates closer together, the balance between solids and air-filled and water-filled pore spaces is dramatically altered. Compaction usually eliminates the largest soil pores first. When a large portion of the initial soil air is forced out of the upper plant root zone, water movement and root penetration are destroyed. Also, compacted soil will tend to hold chemicals at the surface because penetration by water is slower. Discing, deep soil ripping, soil imprinting, and other physical actions can usually remove this problem (Montalvo, McMillan, and Allen 2002).

Salinity in the soil can be common on farmlands, especially in locations where cattle were concentrated, such as holding pens and small pastures. Looking at existing vegetation to determine if there is a salinity problem is not always reliable. Unless you know the age of the existing vegetation, it will be difficult to assess the soil condition because it may have changed after the plants became established and their root systems were lower than the salt penetration.

Salinity is typically expressed as electroconductivity (EC) and is shown as a whole number. For most native vegetation, an EC of 2 or lower is preferred. High salinity will negatively affect seed germination. An EC of 4 will cause plants to show negative signs.

We highly recommended that a soil test be conducted on your site very early in the analysis process. Past usage of the site or adjacent land uses will be a guidepost for how intensive your sampling should be. Contacting local farmers, farmer advisers, agricultural offices, and soil conservation offices should be helpful. At a minimum, testing should be conducted for pH, salinity, EC, percentage of organic matter content, texture, macronutrients and micronutrients, and any other chemicals that are known locally to be of concern. Failure to conduct even the simplest soil test can lead to significant consequences (fig. 7-4).

"...and it will cost 10% less if we don't do soil testing"

FIGURE 7-4. Conducting basic soil sampling for every project site can yield valuable information for design decisions.

Soil depth and texture can vary widely over relatively short distances. For example, alluvial deposits on floodplains and benches adjacent to streams are typically composed of a mosaic of gravel, sand, silt, and clay. Restrictive layers—for example, lenses of clay—can prevent or retard root penetration. Restrictive layers can be a detriment if you are trying to get plant roots to grow to the depth of permanent groundwater, or they can be an asset if you are trying to restore a vernal pool or pothole where the intent is to hold seasonally ponded water. Additionally, lenses of sand or gravel beneath irrigated plants can intercept and conduct water away laterally, preventing roots from growing downward to the water table.

Some restorationists always dig several backhoe trenches on their project site to inspect for variations in the soil profile. (Note: If you dig a trench, be sure to follow Occupational Safety and Health Administration standards for shoring up the sides of the trench to prevent collapse on observers.) If the site was formerly planted with an agricultural crop, it may be possible to identify areas of problem soils by analyzing historic aerial photographs and looking for differences in plant growth (e.g., the health of orchard trees). This type of information can also be obtained by talking with farmers who formerly worked the land.

Soil Preparation and Handling

Earthwork can be very time-consuming and expensive in terms of labor and equipment costs, especially if soil must be hauled and dumped off-site. One of the highest component costs in most wetland restoration projects is earthwork. In most cases, earthwork activities involve mechanized equipment and skilled operators to achieve safe, efficient, and desired results. Earthwork activities for restoration projects vary, from scraping off a thin layer of earth, to removing an undesirable seed bank, lowering the finished grade by a yard or more to establish a soil surface that is much closer to the known water table level, or performing cleanup activities from erosion or sedimentation problems on your site. We encourage you to seek the services of knowledgeable professionals to help you develop an earthwork or grading plan. This will ensure that you maximize the opportunities to balance your earthwork activities to minimize the need for importing or exporting any earth material from your site. The following factors should be considered as you plan earthwork activities:

- Earth moving should be planned to be accomplished in a single operation during optimal weather conditions. Grading during the winter months exposes the soil to rainfall and the high potential for surface erosion and stormwater pollution. Know how long it will take to conduct the grading and the specific measures you will use to prevent excessive erosion. Will the area being graded have suitable topsoil or duff that can be used on other parts of the project or applied to the newly graded surface? If so, then you may need to stockpile the soil until you are ready to apply it to the final contours. When stockpiling the soil, make sure to separate the topsoil from the subsoil (see the discussion on maintaining soil stockpiles at the end of this section).
- Excavated material needs to have a place to be relocated, preferably permanently. The specific location of the receiving site from the source site will affect the cost and time of the operation. In some instances, it may be possible to sell the material and have the buyer

haul it off-site. Whether on-site or off-site, the receiving land should be examined to verify that no sensitive resources will be impacted.

- Importing soil should have the same characteristics as the final destination, unless the change in soil is specifically required by the design. Imported soil may also bring in undesirable organisms, such as weeds and soil fungus. Evaluate the benefits of the soil against the measures that may be needed to control the weeds or other negative aspects of the soil.
- Can you, or your professional or volunteer workers, operate the required equipment?
- Account for paid operator time, typically at an hourly rate. In some cases in which the operator owns the rental equipment, a combined rate may be available. Operators not familiar with ecological restoration need to be made aware of the constraints prior to beginning work. Excavation or fill operations frequently require additional labor for handwork. Excavation of material may leave a substrate condition that is atypical, which may affect the performance of your project. Soil amendments and preparation will require money and time.
- Will local ordinances, regulations, or laws restrict the noise levels of equipment operation and under what specific circumstances? Determine the maximum noise levels that will occur on-site. In some cases, the time of day will determine acceptable noise levels.
- Do you know the specific equipment needed for the job? There are many types and sizes of equipment available, and not all types of equipment do everything your project may require. Interview people who have used the equipment that you are considering for your project.
- If the grading operation is small, can it be done with hand equipment? Consider the number of volunteers and total hours. Are they hard workers? Is the haul of excavated material minimal and suitable for hand equipment? Perhaps simple pickup trucks will be sufficient to haul the material out of the area.
- If it is a large job, will it require excavating-type equipment, such as a bulldozer or backhoe?
- Can the equipment place the excavated material in its final location?
- Will the grading require vehicles to haul material away to a final location, and is it very far away?

In many cases, sites will not require landform modification because they are derelict lands, either from past industrial use or from agricultural practices. Because abandoned lands typically are infested with numerous weeds and with high nutrient loads in the topsoil, essentially no native species are capable of becoming established there. The most frequently used techniques are extensive thatch removal and several applications of herbicide treatment, which greatly reduces the soil seed bank. Another approach involves discing of the soil on a regular basis until the seed bank becomes exhausted. Another method involves "soil inversion," a process of turning over the topsoil and bringing up the subsoil from three feet below the surface. This technique has several advantages over the other techniques because it moves the nutrient-rich topsoil three feet below the surface and brings the subsoil to the top. The nutrient levels of the subsoil will not impede sown native species. In addition, moisture will penetrate more quickly down the soil column.

At times, there will be a need to excavate large holes or trenches that will later be filled and planted. It is critical for successful plant establishment to maintain the soil horizons when refilling excavations. To accomplish this, segregate soils during excavation and provide separate piles for the soil horizons. Stockpiled soil should be clearly defined and identified were it occurs in the soil profile. If the stockpile is to remain for a long time, it is good practice to record location of collection (if not nearby the excavation), fence the stockpile, and provide for erosion control. Re-application should be done in layers in reverse order (reverse backfilling) following the soil profile (making a layer cake).

Erosion Control and Slope Stabilization

In most areas, it is necessary to install temporary erosion control measures wherever disturbed areas will be exposed to wind, rain, or snowmelt erosion during the construction period (including temporarily stockpiled soil). There are many sources of information on erosion control best management practices. Many state and regional water quality control boards have published field manuals for erosion and sediment control.

There are numerous biotechnical strategies and designs for using live and dead native plant materials for simultaneously achieving slope stabilization (e.g., stream bank stabilization), erosion control (e.g., gully control), and native plant community restoration. Useful information on soil bioengineering is available on the Web as well as in many publications describing the design of bioengineering measures along with photographs, diagrams, and drawings depicting their construction (Gray and Sotir 1996; Hoag and Fripp 2002; Schiechtl and Stern 1996; Schiechtl and Stern 1997).

Commonly used in erosion control strategies are straw bales, straw mats, straw rolls, and similar products containing vegetative matter such as coconut fiber (coir). Seed from undesirable species is commonly found in these products; thus, it is very important to specify that these types of materials should be "weed-free." It is not uncommon for invasive weeds to be introduced onto restoration project sites because of a failure to specify "weed-free" erosion control materials.

Chapter 8

Plant Material

There are several options for helping vegetation get established on your restoration site. One of the most common propagule types is seed, which can be used from existing plants on-site or brought in from another location. Equally common are plants grown in containers of numerous sizes and shapes. Many species can be produced vegetatively and afford a quick and efficient means of establishing a large number of plant specimens on-site with low cost in labor and money. Cuttings, rooted cuttings, and rhizomes are commonly used for many species. Translocation or salvaging of specimen plants or plant assemblages provides an excellent opportunity to put mature individuals on-site quickly along with the soil components of invertebrates and microorganisms. As with other areas of restoration, there are no absolutes on approach. Each type of plant material has advantages and disadvantages that need to be considered in the context of your situation with regard to available funds, labor, space, and time.

Seed

Seed provides an excellent way to introduce many species on your project at a fraction of the cost of container plants. Seed use will permit the various species in your project to sort themselves according to microhabitats or through various mechanisms of plant competition. Although valuable to a restoration project, the use of seed can present some challenges. Even then, however, it is typically well worth the effort.

Purchase or Collect?

Commercial seed suppliers and producers typically collect, grow, and harvest a standard list of species and quantities. Some commercial seed providers have readily available common species that are used frequently on restoration and erosion control projects. Some suppliers may have several species, but, because of uncertain demand, the supplies may be limited. It is not uncommon for distinct subspecies to occur in environmentally unique areas (i.e., coastal windswept bluffs and serpentine soil). Exercise caution when ordering these types from providers because they may not be from your area.

Verify the season and year the seed was collected. If the stock is older, verify a retest on germination; if viability is reduced, this will alter your bulk quantities. Federal law requires retesting the germination rate of each seed lot after a specified time. If possible, visit the seed supplier and observe their handling and storing methods. If the supplier is unknown to you, it may be worth the time and cost to have a second test conducted to ensure performance of the seed.

The advantages and disadvantages of using provider-supplied seed are as follows:

Advantages
- Extensive labor is not required to gather seeds.
- Seed is stored by the provider and delivered when you desire.
- Seed is cleaned to standards, yielding higher seed counts per weight.
- Providers typically provide germination and purity rates for each species or lot.
- Some providers can provide advice and make recommendations that can enhance your seed mix (e.g., diversity, germination).
- Providers can supply large quantities of common species for the region.

Disadvantages
- Determining the origin of seed may not be easily done.
- Seed may have unacceptable weedy species also present.
- Seed storage might be an issue with some suppliers.
- Less common species typically are not available; generally, only the most common species are stocked routinely.

For larger projects, notifying seed suppliers and growers well in advance can eliminate supply problems for many species. However, suppliers are typically less willing to speculate on sales of an uncommon or infrequently requested species. Many species have a limited shelf life, and a supplier cannot afford to pay for the collection or grow out of seed that is never sold or that substantially loses its viability when stored for more than one season.

There are two options for obtaining uncommon seeds: (1) negotiate a contract for custom seed collection or (2) establish within the scope of your project a program to collect seeds. The advantages of custom seed collection include all of the advantages of using provider-supplied seed listed earlier. Also, it is possible to get most, if not all, species in the design. The disadvantages of custom seed collection include the following:
- Depending upon species, you may not be able to get the quantity desired.
- For more uncommon species, identification of subspecies and variant may not be correct.
- Typically, custom collection costs significantly more. Harder-to-find species will cost more.
- You must provide storage (a cool, dry environment).
- Not all species will have collectable viable seeds available at the same time of the year.

Collecting Your Own Seed

Seed collection sites should be identified in advance of the collection time. This will allow for obtaining appropriate permissions and permits, if required. When working with threatened or endangered species, have ready all appropriate documentation and permits required by the governmental agencies.

Most important when collecting your own seed is determining when to conduct the harvest. Collecting mature seed is critical to a successful sowing operation; collecting too soon will result in poor germination rates on-site. Visits to the sites prior to collection will help you understand how the plants on the sites are reacting to the environmental conditions. Other sites may not be exhibiting the same development patterns. Locating seeds can be difficult. Suitable sites with the species you need may not be accessible for collection. Public lands, parks, wilderness areas, and other land management organizations have specific procedures on how the resources are to be managed. Private owners should always be contacted prior to collecting on their property. Advantages and disadvantages of collecting your own seed follow:

Advantages
- The seed is ecologically adapted to the local environment of the site.
- Low-cost collection procedures can be used.
- The seed can be collected as needed, meaning no long-term storage requirements.
- No expertise is required to grow plants (in contrast to the seed increase method discussed later).
- A much higher diversity of species is available from the site than is possible from purchased seed.

Disadvantages
- Collecting introduces weedy species.
- Finding donor locations can be difficult.
- The seed will not be appropriate for planting if it is collected from stressed, diseased, or insect-infested plants.
- Species must be accurately identified; variants or subspecies can be an issue.
- Additional time is needed to obtain appropriate permits, clearances, or permission.
- Depletion of the natural stock on the donor site is possible, depending on the efficiency of the collection methods.

Seed collection does not need to be species by species. It is common to apply gross collection techniques in which several species are harvested at the same time, such as mowing prairies and grasslands. The seed and vegetative matter harvested together is often referred to as hay. This hay also provides mulch and microenvironments for promoting seed germination. In the case of projects conducted over a long time span, this method may be preferred because there is no need to clean seed or calculate purity. It does, however, mean that you will not know the quantity of each species being applied. This should not be a problem for multiple-year projects dominated by annuals, because spot seeding can be conducted the following years. As with any seed, the material should be kept dry and cool, stored out of direct sunlight, and protected from potential insect and rodent infestation. Use paper bags (not plastic) for aeration and reducing mold growth.

For collection methods, several trips to the same site will be necessary to obtain the range of genetic types with early and late seed set. Not all species or individuals produce the same quantity of seed from year to year, even with suitable weather. Some species with specific habitat requirements may occur in small populations, requiring that you collect in a significantly large region to obtain sufficient seed without exhausting the donor sites. Know your species and any of its special traits so you can plan for possible backup measures.

One such measure may be to establish a controlled seed bed planting, a method called "seed increase." This method is very effective for providing material for annual and perennial herbaceous species that is otherwise difficult to obtain. A large number of donor plants should be used to maximize genetic diversity. Seed increase is very effective when working with limited natural vegetation or with locally endemic, endangered, or sensitive species. Some advantages and disadvantages of the seed increase method follow:

Advantages
- You can determine the amount of available seed seasonally based on growing beds.
- Controlled seed beds permit a higher purity of seed with much less weed and extraneous vegetative matter.
- Controlled seedbeds with ample nutrients and water can improve the viability of seed collected for most species.
- Seed collected and increased on-site ensures site suitability.

Disadvantages
- It requires space; depending on the number of species, this can be a significant area or represent an added cost for land leases. If done by contract this cost is typically passed along in the cost of the seed.
- It requires at least one season's growth (and possibly more) in advance of the project. If the project is a multiyear installation, then this is not necessarily a disadvantage. Current recommendations are to collect seed for no more than three years, to prevent reducing genetic diversity on the project site.
- Costs are still higher than for stock seed from a provider. If you do seed increase on the project site, the resource cost will be labor, water, and other maintenance costs.
- It requires a reliable source of continual labor to maintain the seedbeds over an extended time without interruption.

Seed increase can be contracted to some growers. Your primary concern in inspecting the operations of the grower is to verify that the facilities are maintained properly and are clean and that the seed will be handled correctly.

Purchasing Seed

Seed is regulated by the government, and a number of requirements are placed on seed providers, such as testing the seed lot for purity and germination. These two measures are important for understanding exactly what you will be purchasing.

Purity

The ability of seed to be separated or collected with a minimum of nonseed material is highly variable. All native seed stock is sold by the bulk pound and typically includes live seed, dead seed, vegetative parts, dirt, sand, and seeds of other species. Most seed suppliers and growers perform standard cleaning operations to remove the nonseed material. However, there is a point where the level of effort exceeds the resulting product

Providers sell nongarden seed by the bulk pound. The cost for a bulk pound includes all those "nonseed" elements listed earlier. Therefore, knowing the purity of the seed will enable you to determine the seed quantity and the cost of that seed being purchased.

Calculating the quantity of seed from the bulk seed poundage is a simple step and is explained in box 8-1.

Germination Rate and Viability

Germination rate and viability are different terms addressing aspects of growth potential in seed stock. Viability describes whether a seed is capable of germinating by establishing the presence of an embryo plant within the seed coat. Germination rate addresses the total number of seeds that actually begin to grow a seedling in a sample. Viability includes germinating seed plus the dormant seed that requires longer to actually germinate.

Knowing the viability of the seed stock is critical for planning a successful project, especially when using rare or endangered species. Several environmental factors contribute to the development of a seed, and many native species produce seeds that appear to be normal externally and internally but that fail to germinate for various reasons, including atypical rainfall at a critical time of seed development, genetics, fungus infection, or insect herbivory, to name just a few. Seed viability can vary from season to season, making it difficult to get the same results. A simple germination test that you can perform involves placing seeds on moist germination paper or a moist paper towel and maintaining them under controlled conditions to promote germination. After about seven to fourteen days, you can calculate the percentage of seeds that have germinated. A more frequent, and quick, test for germination is to apply tetrazolium chloride, although this test is less accurate than a full test for viability.

In some cases, you may want to limit using seeds for testing but may still need to verify seed viability. Radiographs (x-rays) have been used in cases where there are too few seeds to permit using some on testing for viability. The radiograph will detect the presence of the primitive root but cannot predict if it is capable of germination. Radiograph tests can also determine whether embryos are present in seed that takes several years to finally germinate. Seeds have a dormancy period, which for some species can be several years. Germination testing is not practical in these situations. Also, some seeds have a hard coat that can take years to deteriorate. In addition, some species require fairly elaborate or complicated treatments to cause germination, such as scarification, cold storage, stratification, or fire treatment (Leck, Parker, and Simpson 1989). Other species have a very short shelf life.

Certified seed sold in the United States must, by law, display the purity and germination percentage on the seed tag attached to the seed bag. As discussed earlier, more than just seed is included in the bag. Many—but not all—seed providers supply a PLS ("pure live seed") percentage on the tag. Knowing the PLS of your seed lot is an important first step in developing a seed mix that will provide the desired results.

Pure live seed is calculated by multiplying the purity and germination rates by the bulk quantity of seed. Thus, with seed tags containing a purity rating and germination rate, you have the ability to determine the weight of live seed. Knowing the cost for a bulk pound, you can then determine the per pound cost by using the formulas shown in box 8-1.

Box 8-1. Calculating Pure Live Seed and Cost

A seed test report includes the following:

Pure Seed	73 percent
Inert Matter	4 percent
Other Crop	0 percent
Weed Seed	3 percent
Germination	47.31 percent
Dormant Seed	15.03 percent
Total Germination and Dormant Seed	62.34 percent

(some use the phrase "viable seed" for this total)

PLS = purity % × total germination % expressed as decimals (to four places)
PLS = 73% × 62.34% ÷ 100 = 45.51% PLS

For this seed lot, the PLS tells us that for every one bulk pound of live seed, only 45.51 percent is live seed. Or, if you bought 25 pounds bulk, in effect you would be getting only 11.38 pounds of seed capable of producing plants on your project.

What would one pound of live seed cost?
The seller charges $5.75 per pound bulk seed for this species.

PLS cost per pound = bulk price ÷ %PLS
= $5.75 ÷ 45.51%
= $12.63 cost for one pound of PLS seed

Understanding the relationships of germination and purity compared to the unit price of the species will enable you to determine your best options. Low-unit-priced seed may have very low purity or germination rates. The result is seed that may cost ten to fifty times the bulk seed weight price. It is always good to calculate the numbers prior to making decisions. Knowing the bulk seed cost is important, but it is only the beginning; more critical is knowing the germination and purity rate of the seed. This will allow you to determine what the actual cost of the seed will be. It is very enticing to see a bulk pound cost of only two dollars, but after viability is considered, the actual cost could be as high as one hundred dollars per pound!

An annotated seed mix worksheet (appendix 6) is provided to permit loading into standard computer spreadsheet software. This will reduce the potential for miscalculations that can occur when working with several numbers and species. The example in appendix 6 is for applying seed separately, not as a mix, because different areas of application are identified in column K. Projects not purchasing seed but collecting their own should attempt to document in some way the quantities applied, for future reference by others. This information may be qualitative in nature, because the vegetative component of hay can be highly variable, as is the seed production from year to year.

Seed Quantities

A frequently asked question is how to determine the quantity of each species to put in a seed mix and how many pounds or ounces should be in a seed mix per measured area. Unfortunately, the variability between sites, along with numerous other factors, makes it impossible to give a specific answer. This area of specifying restoration work requires some knowledge and experience working with each species under similar conditions. Information on how to guide your decisions will come from different sources.

Inexperienced individuals can get information on quantities and mixes from some suppliers. However, because environmental conditions significantly influence the outcome of the seeding efforts, it is important to know the context and circumstances of any anecdotal information to determine whether your site has similar conditions. Although it requires advance planning, setting up a few test plots sown with varying mix combinations can prove very helpful. The key to compiling a body of knowledge compatible with site-specific conditions lies in learning as much as you can from each project, discussing results along the way with other practitioners.

In dry environments, some species do not germinate until the first or second year following sowing. Without seed pregermination treatment, other species may take several years. Knowing how a species responds to sowing will directly influence how you develop the evaluation criteria for the project. Calculating the seeds per area sown will help you evaluate your efforts. Understanding how a species develops and matures may influence some potential measurements to be used as criteria.

Seed application rates are variable. The vegetation community and the composition greatly influence the quantity and weight applied. Grasslands seed mix (lightweight seeds) will have a much lower overall weight than a scrubland seed mix (large, dense seeds) with the same number of total seeds. The site preparation and how the seed is put onto the project site also affect the quantity needed. Removal of exotics and weeds is critically important because competition for light and water will be a disadvantage to the native species being seeded.

Broadcasting seed has its own set of challenges. Seed comes in a wide range of shapes and sizes and with various vegetative structures to facilitate dispersal or soil penetration. The application on-site, if done by hand, may permit the seed to separate in a mix according to size and weight. Belly grinder (organ grinder) devices and seed drills frequently will clog or distribute seed unevenly. This condition can be controlled simply by using your hand to mix up the seed in the tank while grinding. Another procedure when using large equipment is to add an inert carrier to the seed mix. The carrier should have similar density as the seed to ensure a consistent mix; for this purpose, bran, polenta, rice hulls, sawdust, and even sand have proven successful. The carrier should be kept dry at all times to prevent clogging your hand-operated seed broadcasting equipment.

It may be easier to visualize seed quantities by determining how many seeds you want in a square foot. Consider all of the species you want in your mix and whether you want every species in the same density or not. It is much easier to visualize fifteen seeds in a square foot. Then calculate up to an acre. By back calculating, you can then determine the amount of bulk pounds you need to have a PLS of fifteen seeds in a square foot.

Some restoration designs will include species that function as "nurse plants," nonnative plants that do not persist on-site but that help to nurture slower-growing natives. To be an effective nurse plant, the plant cannot directly compete with native plants in a way that interferes with the establishment of the native species on the site. Competition for resources and persistence on-site long after the initial establishment period are concerns when using nurse plants. Many native plants can serve as effective nurse plants. As you study various reference sites and observe the interactions, you will develop an understanding of how to adjust quantities and select the species to get your desired results.

In projects with several very limited species, or for which the seed is extremely hard to obtain, application of the seed may be altered. To yield more predictable results, follow carefully controlled application techniques in precise, suitable locations to plant the "very rare" or "very expensive" seeds.

Seed Services

One locale can hold several seed suppliers, with a number of them offering additional services beyond providing seed. Some governments also have seed services (e.g., plant materials centers) that may be available under specific circumstances. The potential range of services includes the following:

- Assessment of the maturity of seed for collection (including quick delivery of a sample to the lab for assessment; arrange in advance to ensure service)
- Evaluation of the seed lot for debris amount
- Cleaning services for various species (this will increase the purity of the seed lot)
- Advance assessment techniques, such as radiographs to determine viable embryos
- Cleaning of empty or damaged seeds (further purification of the seed lot)

- Final determination of purity
- Measuring moisture content of seed (requires using a small amount of seed)
- Storage facilities available for rent
- Germination tests (taking up to four hundred seeds)
- Cut test on fifty to one hundred seeds (can be used on all seeds, even seeds less than 1 millimeter, because the x-ray test is unable to differentiate structures on smaller seeds; the cut test, as with the radiograph method, determines the presence of an embryo in the seed)
- Seed drying to reduce moisture content of the seed to enhance long-term storage

Note, however, that not all providers or labs offer all of the services listed here.

Seed laboratories and businesses operate differently throughout the world. Prior to using a local provider's services, verify the services, conditions of service, product provided, time, basis of cost, and other contingencies.

Seed Collection Guidelines for Native Plants

As discussed in chapter 5, species selected for a restoration site are primarily directed by the vegetation community, location, slope, aspect, and soil conditions. However, even when your species list contains appropriate species for a specific site, it still may not be feasible to obtain specimens of all species occurring on-site. Nurseries have not propagated all species that would be desirable in a project, because either the species are rarely requested or the methods of propagation have not been resolved.

In recent years, various conservation organizations and government agencies have expressed concerns about the overcollection of some native materials and the genetic suitability and diversity of the material collected. Some of these organizations have developed very detailed lists of criteria for collection and documentation of seed. The following list summarizes the most important issues raised by these organizations:

- Obtain permits (if required by government agencies) and permission from the property owner prior to collecting plant materials. Species in wetlands and those that have special designated status (e.g., as endangered, at risk, or sensitive) or cultural significance to indigenous peoples require special attention.
- Give priority to collection sites as close to your project as possible.
- If infeasible to collect nearby, select sites with similar environmental conditions, rainfall, temperature regime, elevation, soil conditions, aspect, and any other factors that may influence the vegetation you are restoring. Collect from sites in the same watershed if they meet the aforementioned conditions.
- Be sure of your identification. There are several species that have subspecies and variants with dramatically different physical forms. There are also species that are very similar to others but have significantly different habits that may not be suitable for your project.
- Avoid collecting from isolated stands or individuals outside of a main concentration of plants. Also avoid unhealthy or atypical plants, which will reduce the problem of collecting genetically inferior plants.

- Collect from at least fifty individual donor plants of the same species. Try to collect similar quantities from each donor plant. This method will avoid emphasizing a specific genetic type.
- Pick colonies that are widely spaced from one another to avoid collecting from clones or individuals with the same genetic composition.
- Do not collect from plants near or part of the general landscaping; these plants may not be local but instead provided by a nursery with an unknown origin or contaminated genetics.
- Collection quantities vary depending on the habits of the plants. Many experts recommend not collecting more than 10 percent of the seed produced in any year. This is critical for annual plants that rely on the seed crop for the following year. Shrub and tree species produce vast quantities of seed that do not germinate and that become part of the seed bank or are eaten by various animals. Collecting higher amounts of these would not necessarily be unacceptable, although there may be some species in this category that would warrant special treatment.
- Always fill out the seed collection forms and attach them to the bags while out in the field to avoid labeling errors. Be sure to enter all appropriate data into your plant material records to ensure source origins, collection dates, and quantities. This will be an important source of information later if these plants are propagated and used on other sites or areas of your property.

As these guidelines are under continual revision, we suggest you refer to appropriate organizations for the most current guidelines should you need to collect seed.

We cannot overemphasize that location and distance from your site can affect on your project. The question of "How far can I go?" is difficult to answer. Many organizations and some public agencies require material to be collected on-site or, at a maximum, within the watershed where the project is located. In places of the world with significant topographic relief, a watershed may be only a few hundred hectares.

Very few studies have concentrated on non-timber-related native species. Preliminary studies with some coastal sage species in southern California indicate visible differences between populations about fifty miles apart (Montalvo and Ellstrand 2000). The effect of moving populations from one area to another is not fully understood. A judgment has to be made on distance. As in the list above, we suggest giving priority to the collection sites as close to your site as possible.

Other Types of Plant Propagules

Various other types of plant propagules can be collected and either installed directly onto your site or "potted" and grown to adult size in a nursery. Table 8-1 compares the various aspects to consider for the several plant material forms possible for restoration projects.

Plugs

Commonly used by producers involved with seed increase or contract grow operations, these plants have a small but developed root system that enables handling on-site for immediate plant-

Table 8-1. Comparison of Various Plant Propagules Used in Restoration Construction Projects

Material	Requirements	Constraints	Planning Obligations	Operation Obligations	Issues and Solutions
Seeds	Suitable surface substrate.	Application at specific time of year.	Supply issues may require up to two years advance request.	Moisture and soil environment critical for germination. Some seed require special pretreatment.	Seed should be collected in a defined geographic area for ecotype considerations.
Cuttings	Collection during dormancy of plant.	Holding of cuttings requires cooling or refrigeration.	Schedule of project timed so collection and planting occur in a very short time period (days).	Prevent drying out prior to leaf sprouting. Improved survivorship attained by soaking in water for one day prior to planting.	Cut planting end at angle, opposite end square, to avoid problems. Cuttings can dry out.
Rooted Cuttings	Place cutting in water or wet burlap for root growth.	Time of year important; rooting allows for timing flexibility.	Allows for a little more time flexibility in scheduling.	Planting of rooted cuttings should proceed with little delay.	Roots may be tangled or curled depending on storage. Wet burlap helps in handling material. Cut any roots showing signs of twisting or curling.
Tubes or Bands	Requires careful handling.	Extra care in planting to avoid breaking roots. Wet soil to prevent drying of root ball.	Short holding time before planting.	When delivered to site, should be maintained closely. Water to prevent drying while waiting to plant. Do not lay out in sun during planting operation.	Root ball breakage is primary concern, followed by proper moisture control during planting operations.
Containers of Various Sizes	Permits planting at any time of year.	Properly prepared planting hole. Plant at correct level with ground.	Requires nine to thirty-six months to attain desired planting size.	Moist planting holes; backfilled completely with no air spaces. Requires supplemental irrigation.	Delayed planting may cause severe plant damage. Improper handling can break root ball during planting operation.
Pole Cuttings	Deep groundwater; no supplemental water. Poles 2 to 4 inches in diameter and 4 to 8 feet long.	Should be collected in dormant or near dormant time of year.	As above for other cuttings.	Poles should be deep enough to reach the moist soil zone.	Collect only a few poles from any one plant.
Salvage Translocation	Need for large specimens on-site for fast habitat development.	Equipment may require large area or have slope limits and cause damage to other plants.	If large equipment required, plan its presence on-site prior to smaller plantings. Delay in other plantings will alter normal planting schedules.	Usually requires supplemental irrigation immediately.	Plants taken within groves may not have sufficient reactionary wood to be self-supporting.
Sod or Vegetation Blocks	Need to cover area with several species not commercially available.	Rooting depth, size of block, and weight will determine equipment.	Identify areas in advance of salvage material location and quantity expected. Prepare donor site and receiving site appropriately (wet receiving site).	Placement is important. Sides of blocks must be flush with another block or filled with soil if planted singularly. Water following placement to encourage soil-root contact and reduce air pockets.	Donor site not always available. Method is excellent for introducing several soil organisms, and herbaceous species not otherwise available.

FIGURE 8-1. Plastic tube propagation in flats; from here, depending on species, the plants are planted at sites or put into larger containers. Tree of Life Nursery, San Juan Capistrano, California. (Photo by John Rieger.)

ing. A dibble or similar device is used to provide a small planting hole. The advantages of using plugs are the quantity that can be handled at one time (fig. 8-1) and the number that can be planted per given time. Also, plugs have a wider planting window than seeds, using plugs permits precise spacing, and the plants can produce seed in the first season following planting. Many plants are grown in this manner.

Prevegetated mats are used in areas where erosion control and soil stabilization are concerns. Mats are also used where unstable substrates, waves, and currents limit the establishment of aquatic vegetation. They are composed either of a plant mix grown on a soil medium that is in the top layer of a netting material or of plant transplants that are inserted into the soil medium in the upper layer of the mat. The mat is generally made of polypropylene and plant fiber layers but may also be constructed entirely of fiber (e.g., coconut fiber mats). Prevegetated mats have been used to establish submerged aquatic vegetation (Boustany 2003) and wetland vegetation.

Stem Cuttings

Many species of plants can be propagated by stem cuttings (i.e., vegetative propagation). Some species require more care because the growing tips of stems are not always amenable to cutting year-round. These cuttings can be placed directly on-site, as is the case with several willow species (*Salix* spp.). Other species are not as hardy and require growing in small containers, such as four-inch pots, tubes, or bands (fig. 8-2), before being able to withstand the conditions of being planted on-site. Root growth pattern can also determine the type of container. Unlike the standard nursery trade for exotics, in which containers are sized and standardized, many native species are not so easily classified. For many species occurring in coastal North America, the container used for nut trees has been successful in training roots to grow down and limit the amount of coiling.

Although planting species at the smallest size possible is highly recommended to permit the plant to adapt immediately to the site conditions, it may be necessary to plant larger-sized containers. A larger root ball can hold more moisture and is more resistant to drying out. If getting water on-site is an issue, this may help alleviate it. Larger specimens make an immediate impact and can be seen from a distance. Stakeholders commonly require some visual stimulus to be convinced that the site is successful. This concern is part of the politics of restoration projects, and you must carefully evaluate how to commit the project funds and keep your stakeholders satisfied, not always an easy situation.

Wetland plants initially were difficult to obtain; however, in some regions, numerous species are now available as the restoration practice continues to expand. Cost may be a significant factor in obtaining sufficient numbers of infrequently requested species. Rooted woody cuttings are another way to get specimens on-site without relying on seed availability. Collection of cuttings from a donor grove can be a quick and cost-effective way to get container plants started. Several non-

FIGURE 8-2. Container production of several species at the National Wildflower Centre, Landlife, Liverpool, England. (Photo by Mary F. Platter-Rieger.)

FIGURE 8-3. Cottonwood and willow container production using cuttings from local donor plants. The drip irrigation system provides water to each of the several thousand containers in this cost-effective greenhouse. The Nature Conservancy, South Fork of Kern River, California. (Photo by John Rieger.)

profit groups have developed a system of small greenhouses with containers and cuttings supplied with water (fig. 8-3). These are then planted like a regular container plant at the best time of the year. The containers are cleaned and then reused the following year, maximizing the economical and efficient nature of this approach.

Bare root plants have been grown in a modified soil bed that permits removal without extensive damage to the major root system. This technique can be used only during limited times of the year because the plant is vulnerable to desiccation. For one major project, the vendor developed a technique to cut the willows out of the ground, encase the roots in a polymer, and refrigerate until planting. Several thousand plants were successfully installed using this method. The operation required coordination with the nursery and the construction schedule to provide a place to put the plants at the right time of the year. The rooted cuttings were started before the contractor was hired to prepare the planting site. Operations like this require careful scheduling and communication among the various team members.

Seedlings are usually plants grown from seed directly in plugs or small pots. To avoid investing in large containers and the required soil mix, planting of seedlings can be another approach to maximize your resources.

Long woody cuttings have been used in areas where cattle may browse on lower hanging branches. Used in the southwestern United States along streams, creek beds, and small ponds, the long woody cuttings are the same as the whips, or live stakes. In this case, the cuttings can be between four and ten feet long.

Container Plants

The term *container plant* is commonly applied to specimens grown to the "one gallon" size or larger. Projects requiring container plants will have numerous choices to make concerning the size of the plant and the shape of the container. The size of the plant to be used is based on several factors previously discussed. Plants provided by typical commercial nursery operations will most likely be propagated in one of the traditional types of cylindrical containers. Following are some advantages and disadvantages of ordering plants from nurseries:

Advantages
- Do not need to know propagation techniques
- Do not need nursery grounds taking up space
- Do not need to maintain support facilities and staff for plant propagation
- Delivered to site when ordered
- Can control the quality of plants provided on-site by refusing those not meeting specifications
- With advance notification, can order almost any quantity desired (may require contract growing arrangements)
- Not involved with locating a wild site or obtaining required permits or permission

Disadvantages
- Delays caused by the site not being ready may cause deterioration of plant quality.
- Immediate expenditure of funds requires controlled budgeting.
- Plants may not be from immediate locality (if that is a concern).
- A delivery charge may offset savings.
- A large number delivered at once will require a well-organized labor pool to install plant material.
- Provisions are needed for short-term maintenance while plants are installed.
- Container-grown plants commonly allow plants with poorly adapted genotypes to survive.
- Container soil is commonly overfertilized and roots keep exploiting the potting soil instead of the native soil on-site.
- Holophytes (mangroves) grown in nurseries many not be exposed to salt so will need brackish watering prior to planting.

In the United States, plants generally come in sizes referred to as one-gallon, five-gallon, and fifteen-gallon cans (table 8-2). However, no standardization of sizes and dimensions for those container sizes appears to exist. Many nurseries specializing in native plants are now using differently shaped pots to grow various native plants, responding to the specific needs of the species. This is a significant step toward ensuring good, healthy plant material.

Plants with strong, deep roots or taproot-like systems do better in longer, narrow containers than in shorter, cylindrical pots (Burkhart 2006). Plants with fibrous network root systems do not generally need deep pots in the nursery environment and can do well with the standard sizes and shapes of commercial production.

Most pots are made of plastic and paper pulp. Although common in the past, metal cans now have very limited uses because other materials perform better. Reusable bands are made of paper and plastic. Small four-inch pots are made of plastic and are commonly reused.

Except for larger sizes, most production nursery plants are meant to be planted in one to two growing seasons. Larger plant sizes and plants held in pots for a longer time than planned will require larger containers when potted and possibly containers made of different material to withstand the nursery environment for this extended time period. If you have ordered plants in paper

Table 8-2. Plant Propagation Containers

Group	Sizes	Comments
Plugs	1 × 1 × 2 inches	Grown in trays and removed to direct planting on project site or seed increase operations
Paper pot	1.5 × 1.5 × 6 inches 2 × 2 × 6 inches	Pot lasting up to one year in nursery; held in plastic trays
Tubes: leach, super	0.4–0.6 × 0.75–8.5 inches	Reusable pot; slow- and fast-growing species
Small pot: four-inch pot	3.5 × 3.5 × 3.25 inches	Used for herbaceous perennials; used in irrigated areas due to shallow roots
Plant band; variable sizes, liner	2 × 2 × 7 inches tar paper typical 1.5 × 1.5 × 10 inches	Use with box to transport; one-year life in nursery; homemade; removed on planting
One gallon: can*	6-inch radius × 7 inches to 1.5 × 1.5 × 14 inches	Typically for woody tree or shrub; plastic pots reusable
Five gallon: can* Fifteen gallon: can* and larger	10.25-inch radius × 12 inches 18 × 16 inches	Trees and shrubs

*This is a typical example. Check with providers or manufacturers for other dimensions because there are numerous combinations.

pots or tubes, it is important for the health of the plants to be on schedule to receive the shipment from the nursery. This applies to your own nursery operation as well. Plants held longer than planned can become damaged in the smaller container, or the container may deteriorate and not be suitable for handling (box 8-2). In each instance, this will mean a cost to you. If you absolutely need the plants, they can be sized up; typically, this will cost extra, which can seriously affect the project budget.

Even if sizing up plants is not an issue, you still may face increased costs depending on the contract with the nursery. Some nurseries are confined by space and may require a monthly charge for holding plants longer than their established production schedule. Having plants held in the nursery will restrict the nursery from propagating their next crop.

Although requiring a lot of material and labor, operating your own nursery for your project may be viable. Projects requiring several years of planting and large numbers would be excellently suited to this approach. Volunteers have been instrumental in making nurseries function for many different types of projects, from city parks to national parks. The nursery workers at Joshua Tree National Park, for example, were able to develop a creative solution to meet the planting challenges of the Mojave Desert and the growing characteristics of desert plants. The containers they use are actually large, six-inch-diameter PVC pipe that is thirty-two inches long, called tall pots (Bainbridge 2007). The roots grow down quickly and are able to provide sufficient moisture for the aboveground plant to thrive in the extreme desert conditions.

If you are operating your own nursery, your costs will include not only the monetary cost of the new pot but also the labor costs for sizing up and maintenance. Advantages and disadvantages of operating your own nursery include the following:

Advantages
- Can replant or size up plants if planting is delayed, to prevent causing root-bound plants
- Additional cost for sizing up will be only for supplies and labor.
- Can adapt schedule and have stock ready earlier in season
- Overall costs may be lower if handled on own.
- Can change quantity needs without contract modifications
- Can control the quantity and quality of uncommon species
- Will know precisely where stock material was collected

Disadvantages
- Requires a commitment of space for propagations and support of the plants
- Supplies and labor required consistently for maintenance
- A constant drain on the budget (in addition to the initial set-up costs)
- Some species may be difficult to propagate or beyond the knowledge of individuals.
- Coordination and reliability of labor pool (if using volunteers) may not be consistent.

Box 8-2. Lesson Learned: Inspection of Plant Materials

A restoration contractor was asked to hold a quantity of contract-grown oak seedlings over for another year because the restoration site was not ready for installation. Since the crop of acorns was poor that fall, the decision was made to have the native plant nursery that had collected the acorns and grown the seedlings keep the plants until the following spring. The nursery repotted the oak seedlings into larger containers; however, the restoration contractor failed to inspect the plant materials before they were transferred into the larger containers. When the young oaks were delivered to the project site the following spring, they were inspected, looked fine, and were installed. By the following summer, many of the young trees had died even though they had adequate irrigation. When the trees were unearthed, it was found that the roots had encircled on themselves in the original containers prior to being transplanted into the larger containers. Subsequently, the client made the contractor arrange for new, healthy oak seedlings and pay for both their growing and installation.

The contractor learned that it is important to inspect plant materials at the nursery at various stages in the growing process.

Specimen Plant Translocation and Salvage

The option of using well-established vegetation or specimens may become available. The possibility may arise on projects with remnants of established vegetation or specimen plants (chapter 10). Typically in such cases, there will be small clusters of trees or shrubs on properties that have been avoided or where clearing was not required. These can be located between fences, on extreme edges of property, or in odd-shaped parcels of little use to the landowners. Collection of these specimens can bring immediate added value to a project.

CANDIDATE SPECIMEN SITUATIONS

Evaluating the candidate specimens is critical for successful translocation. Situations that may have candidate specimens include the following:
- Isolates left from previous construction activity
- Patches between buildings, roads, or similar structures
- Adjacent or nearby property where vegetation can be removed with owner consent
- Existing material on-site that needs to be removed
- Donor site where material can be harvested periodically
- At or near outlets or culverts where continued growth may cause blockage of water flow
- Patches between railroad tracks, areas protected by fencing (such as roadsides or abandoned cemeteries), and areas adjacent to agricultural fields that are not tillable by standard farm equipment
- Sites slated for development

Regarding this last point, sometimes suitable plant materials can be salvaged and transplanted from a nearby site that will be developed in the near future. Some organizations have established native plant salvage nurseries using volunteers to dig up and pot native plants from proposed project sites with the permission of the project developer and property owner. These plants are later made available to native plant revegetation projects at a nominal cost.

Planning for Salvaged Plant Material

It should be determined early in the design stage whether salvage plants are going to be a part of the project. In colder latitudes, it is not feasible to salvage plants during the winter when the ground is frozen. Therefore, it is important to schedule the constraints into the overall project schedule during the design stage. This will prevent unnecessary delays during the implementation phase. It is important to have a good understanding of when a site will be ready to receive salvaged material and when that material has to be removed from the donor site.

In some cases, you will not have an option of waiting. Providing an interim holding location for the transplant specimens may be an option. However, doing this can be stressful to the plants, so you need to evaluate the duration between translocation operations to allow sufficient time for the plants to recover from the shock of the move. Several factors and conditions must be evaluated to transplant specimens from donor sites. First, the soil is similar to the donor site and in the case of wetlands has the same physical characteristics. Also, supplemental watering should occur immediately after placement. Watering will help settle the soil around the root ball to minimize air pockets. Finally, in the case of moving succulents such as cactus, a holding period aboveground to permit broken roots to callous is required prior to placing succulents in the ground. This will greatly reduce the incidence of bacterial infection through the roots. Some advantages and disadvantages of translocation plant materials follow:

Advantages
- Introduces mature specimens on-site immediately
- Significantly augments visual aspects of the project
- Introduces a higher species diversity than is typically possible with seeding and planting from nursery stocks (this is especially effective with marshes, grasses, and many understory plants of riparian systems)
- Relocated root systems with soil around the plants will also contain numerous organisms that can colonize the site typically faster than if emigrating on-site from adjacent areas.

Disadvantages
- May require unique or larger equipment than is typically present on a project site
- Requires conditions to be successful (to get the equipment close to the plant, there must be a clear area around the specimen or at least on one side)
- Depending upon the equipment needed, the cost can be high and can require adjusting the design approach.
- Ground must be sufficiently firm to support the equipment used to translocate the plant.

- Presence of exotic species in association with the desired plant will require control.
- Clearing or pruning of shrubs and other small plants may be needed.
- Some equipment may have difficulty operating on slopes.

Regarding the last point, be sure to determine the slope and the acceptable equipment access route. You may need to bring the equipment operator on-site in advance to confirm the suitability of access and operating space as well as the number of plants that are salvageable.

Project Implementation

Part III refines your project plans and takes them into reality with the implementation. Whether your project strategy is management oriented or construction or installation oriented, you need to provide a carefully thought out document that explains what is to be done. Circumstances change over time; changes may be necessary, and informing those who come after you will greatly assist in the long-term management of your project.

Despite the fact that a lot of standard construction methods—such as earth moving, planting, and sowing—are incorporated into a restoration project, your goal is not the same as an actual engineering or landscaping construction job. Slopes do not need to be billiard table smooth and flat, and plantings need not be in straight lines or evenly spaced. You can convey most of the desired look in your document plans and specifications, but you will also need to have site inspection. Depending on who is actually doing the work, you may need to train the workers to understand the circumstances they are working in as opposed to traditional construction-type projects.

Chapter 9

Restoration Project Documents

When the scale of a restoration project requires that you work with and through others, you need to develop instructions to guide those who are conducting the work on your behalf. How well you communicate to the project team and stakeholders is by far the most critical element for successful project completion. On many restoration projects, we find that there can be as many as six core team members involved in the planning and design phases, with the possibility of two dozen additional volunteers and tradespeople participating during the implementation phase. Maintaining clear, consistent, and meaningful communications across the large number of project participants and the vast number of activities involved in completing a project is a full-time job.

Architects, designers, engineers, and landscape architects rely on drawings, the bill of materials, and specifications to communicate desired outcomes during the implementation phase of the project. These project documents, sometimes referred to as the "bid package" or "contract documents," are essential communication tools that every design professional relies on to ensure that the project is properly constructed as designed. Restoration projects range in size and complexity, with different needs depending on the purpose of the project, specific site circumstances, and local governmental requirements. Landscape architects and civil engineers are licensed professionals with the appropriate background and training to prepare effective contract documents for ecological restoration projects. We have found that engineers and landscape architects are extremely helpful in accomplishing specific restoration tasks but that the overall project management and coordination must lie in the hands of the ecological restoration practitioner. However, in most cases in the United States for which project construction and installation must go out to bid, either a licensed landscape architect or a licensed civil engineer will need to stamp and sign the construction drawings.

Providing a variety of useful functions, project documents detail in writing the decisions you make throughout the design development process. They communicate your intentions and are frequently required by regulatory agencies during permit review and approval processes. Sponsoring organizations often require that some form of project documentation be developed to ensure that project milestones and objectives are being met. If you use professional contractors to implement any part of your project, these documents are the generally accepted industry tools through which you instruct the contractor and measure contract performance. And finally, project documents serve as key instruments to help you and the project team clearly communicate your planned actions during the project implementation phase.

Project documents typically are combined with specifications and together form the basis of contract documents. As such, these can be legally binding documents and are subject to much scrutiny. We urge you to seek the advice and assistance of licensed and bonded professional landscape architects and civil engineers to help you develop and prepare project documents, particularly if you envision using contractors to implement your project. Contracting with subcontractors or for other services may create potential liability, and the quality of the documents will certainly help you achieve the product you desire. The costs of dealing with errors and omissions on the documents, and the legal exposure that you face when entering into contracts for project construction, are daunting. Again, the scale and complexity of a project should guide your decisions on whether to use professionals when preparing your restoration project documents.

In this chapter, we discuss the three components of project documents: the project plans, the bill of materials, and specifications. Each is a unique element in the document package and has a specific function and purpose for a successful project.

Project Plans

Project plans consist of various drawings that, when combined with the bill of materials and the specifications, provide detailed spatial instructions to the persons responsible for installation. Project plans for ecological restoration projects can have as many as seven types of plans:

- Plot plan
- Removal plan
- Grading plan
- Planting plan
- Irrigation plan
- Utility plan
- Construction details

Not all projects are this complex and, in many cases, some of these plan types may not be required.

Plot Plan

Long before any field data are collected or drawings are prepared, the project's legal boundaries must be firmly and accurately established. Projects occurring completely within the boundaries of a facility, such as a park, government land, or an existing preserve, may have limited concern about the title of the property. However, for a project situated near the boundaries of one or more property owners or where there are known or suspected utilities or access easements, it is abso-

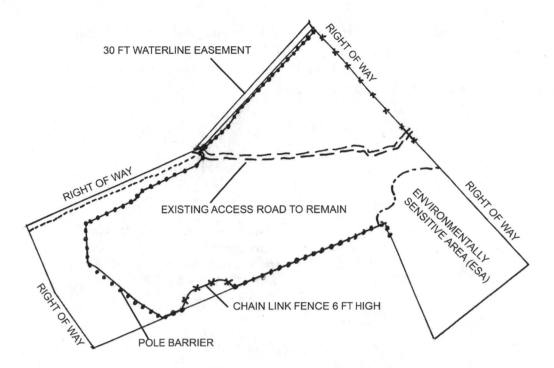

FIGURE 9-1. Project site boundaries indicating property description, legal access requirements, perimeter protection, and exclusion areas (environmentally sensitive areas; see box 9-1). Carlsbad, California.

lutely essential to establish the boundaries of the property that will be affected by the restoration project (fig. 9-1). In these situations, it is a crucial step in the plan preparation process to hire a title company to conduct record searches so that any underlying title issues or previous land use issues are uncovered. If any rights or restrictions are found in the title study, then a licensed land surveyor should be retained to resolve these issues by determining the legal property boundaries on the plot plan. The field-verified plot plan then becomes the basis for all other plans necessary in project implementation. This process is typically needed on recently acquired properties, for which land ownership and rights may not yet have been clearly established (as in the earlier example).

Removal Plan

The removal plan, sometimes referred to as the demolition plan or the clearing plan, provides instructions as to what existing features are to be removed or cleared from the project site (fig. 9-2). Invasive nonindigenous vegetation, dilapidated structures, abandoned utility equipment, trash, and debris are typically identified and targeted for removal and safe disposal at an approved off-site facility. Removal plans document existing aboveground structures and vegetation that are required to be removed. On the other hand, removal plans also identify those elements or features that are to be protected or treated in a specified manner. Unique or sensitive areas, such as drainages, ponds, and areas with rare or endangered plants, are often called out with special considerations. Stands of trees and other vegetation that are to be retained and protected during construction are also shown on this plan. In addition, removal plans clearly identify the legal project boundaries, deed restrictions, and utility easements and call attention to any special site-access requirements that are to be followed during construction.

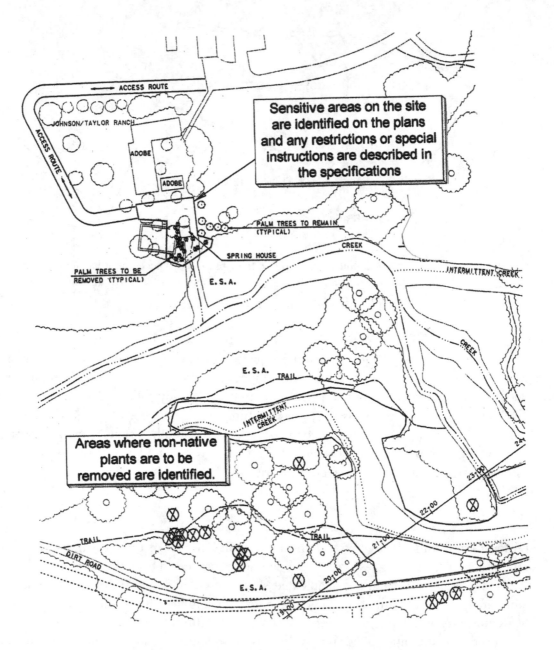

FIGURE 9-2. Sensitive areas identified on a plot plan. San Diego, California.

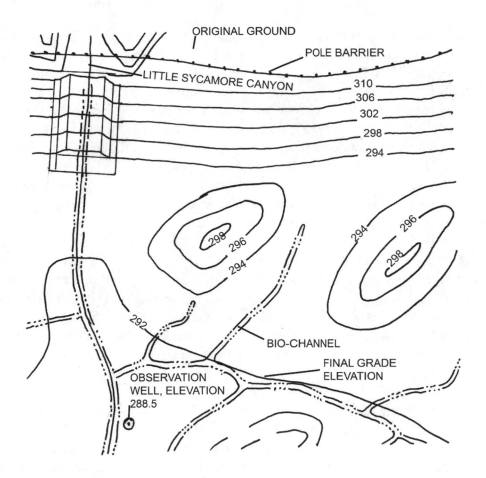

FIGURE 9-3. Portion of the Mission Trail restoration project grading plan indicating biochannels used to concentrate water following flood flows to prevent large ponded areas. Also shown is the slope created by the elevation change and island mounds, which will serve as a refuge during lower flood events.

Grading Plan

Grading plans, commonly referred to as earthwork plans, document what is to occur to the ground surface as a result of implementing your project plan (fig. 9-3). Any manipulation of the ground plane—whether realigning the course of a streambed or recontouring the ground plane to remove a thin layer of soil containing a seed bank of weedy species—requires the development of grading plans. Significant data collection and engineering efforts commonly are required to prepare earthwork plans. Again, the need for professional assistance is related to the size and complexity of the project and the potential for liability, especially from adjacent landowners. In many cases, permission to conduct a project on government land comes with the condition that the plans be prepared by licensed professionals. On other lands, generally the government of jurisdiction will require the issuance of a grading permit, which requires a licensed professional's signature on a grading plan.

Existing natural features, such as rock outcroppings, streambeds, ponds, and even large trees, are identified and carefully plotted on grading plans, as are elements of the built environment, including fences, walls, and structures, such as buildings, drains, culverts, and roads.

Most licensed land surveyors now collect field data with Global Positioning System (GPS) devices, which increase accuracy and reduce time and effort in the collection phase. The electronic spatial data resulting from the GPS field survey is then processed in the office and translated into a base map. This highly accurate base map then becomes the basis for the grading design.

The decision to proceed with more detailed mapping than the GPS base map may be related to the overall restoration program being contemplated. For example, newly acquired land in the beginning stages will want to have the most usable topographic documents so that the planning of the complete program can be done with certainty.

It is during the preparation of the grading plan that the "baseline," or traverse, or reference line, is established on the project site. The baseline is a line placed on the project site over a specific location; this line consists of two distinct end points, with distinct units of measure marked in the field. We have found—after repeated difficulty with finding suitable and reliable points of reference in the field—that a baseline is extremely valuable. First identified on the grading plan, the baseline is then established in the field by the land surveyor. Only the end points and periodic units are marked in the field, with surveying stakes as markers. These markers then become extremely useful reference points for all project participants, from project development through installation and even during the monitoring periods.

As such, we routinely first establish the baseline on the project so that all other features and project elements are being referenced to the baseline. This is helpful when attempting to locate obscure field elements, such as drains or very small plants. The baseline is also a handy reference tool for the team responsible for project construction. Here, the baseline becomes the reference point on the project for where plants are placed and where monitoring sites are established. The units of measure selected depend on the size of your project. For small projects, baselines that

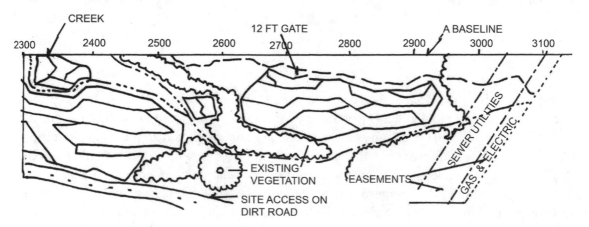

FIGURE 9-4. Establishing a baseline on a project facilitates communication among the workers and ensures the project is installed as designed.

run hundreds of feet in length are common, with every ten to twenty-five feet along the baseline marked in the field. For larger-scale projects, baselines might be thousands of feet in length, with fifty-foot intervals (fig. 9-4).

Planting Plans

Planting plans are used to identify all the species that are to be planted on your project site and to define the spatial relationships between plantings and between assemblages or groupings of species. The amount of detail you show on the planting plan is determined by the level of detail required for actual installation on the site. We have found that the level of detail required for planting plans is far less than many practitioners and permitting agencies realize. There is a cost-benefit ratio that must be considered when deciding the level of detail to include in the planting plan.

Many restoration planting plans contain too much detail, in which the location of each individual plant is shown on the plan even though that individual plant is part of a larger grouping of the same species or part of a repetitive pattern of plantings. Too much detail can translate into confusion in the field during implementation and can lead to questions and undue time spent on plant placement when that level of detail is not warranted. Most restoration projects are not garden designs focused on the placement of individual specimens. Indeed, most restoration projects consist of planting rather large assemblages of the same or similar species. In these and most cases, time-saving drawing techniques are recommended. Figures 9-5 and 9-6 depict the placement of several riparian species for two projects. Notice the use of species callouts—with each callout con-

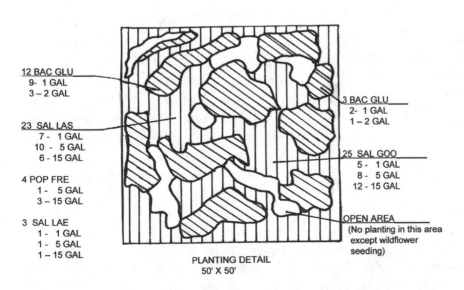

FIGURE 9-5. Planting template for Mission Trails project derived from data collected for the endangered least Bell's vireo. Template is marked off of a baseline, and plants are placed on-site. Each area has species and quantities identified to ensure that the habitat requirements of this endangered vireo are met.

12 BAC GLU
9- 1 GAL
3 – 2 GAL

23 SAL LAS
7 - 1 GAL
10 - 5 GAL
6 - 15 GAL

4 POP FRE
1 - 5 GAL
3 – 15 GAL

3 SAL LAE
1 - 1 GAL
1 - 5 GAL
1 – 15 GAL

3 BAC GLU
2- 1 GAL
1 – 2 GAL

25 SAL GOO
5 - 1 GAL
8 - 5 GAL
12 - 15 GAL

OPEN AREA
(No planting in this area
except wildflower
seeding)

PLANTING DETAIL
50' X 50'

Notes:
1. Transplants are shown on 100' x 100' cell only.
2. The 100'x100' cell is to be comprised of 4 identical
   50'x50' planting cells, with the quantities of plants
   as shown.

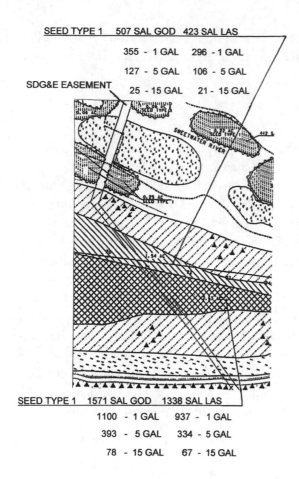

SEED TYPE 1   507 SAL GOD   423 SAL LAS

355 - 1 GAL   296 - 1 GAL

127 - 5 GAL   106 - 5 GAL

SDG&E EASEMENT

25 - 15 GAL   21 - 15 GAL

SEED TYPE 1   1571 SAL GOD   1338 SAL LAS

1100 - 1 GAL   937 - 1 GAL

393 - 5 GAL   334 - 5 GAL

78 - 15 GAL   67 - 15 GAL

FIGURE 9-6. Planting polygons showing different densities of plantings. Less rigorous than shown in figure 9-5, this was a field trial on the effect of two planting densities on growth rate and structure. The triangles indicate large container (thirty-six-inch box) sycamore plantings.

taining the species, size of the plant container, and quantity to be planted—to indicate the area in which the species are to be planted. Although this technique may appear to result in random plantings, it actually is a very structured technique that results in efficient use of time for both the designer and the installing party.

A companion to the planting plan is the plant and planting specifications. This table contains the list of each and every species to be planted on the project. Both botanical and common names are used. Specific requirements for each species, including how each is to be planted and how each is to be associated with other species and sizes, are spelled out on the plant list specifications. When combined with the planting plan, this table (shown in appendix 7) ensures that the detailed intent of the designer is carried out in the field.

Irrigation Plan

In the arid areas of the world, almost every restoration project requires some form of supplemental water during the early stages of plant establishment. Conveying supplemental water to a restoration project is addressed in the irrigation plan. The irrigation plan identifies the source of water, states whether it is potable or nonpotable, identifies how it will be transported from its source to the planting areas, shows the schematic or specific layout of the water delivery system, and specifies the watering regime that is to be followed during the plant establishment phase.

Irrigation plans are routinely designed and developed by landscape architects or irrigation designers. These professionals possess the required skills to develop a design that effectively and efficiently conveys water to plants in their early growth stage. Some project designers install piping and sprinklers or drip emitters, while others believe hand watering for a period of months following installation is more resource efficient though labor intensive. Volunteer participation is critical to make hand watering cost-effective. Irrigation systems come in all shapes and sizes and are determined by your project goals and objectives and controlled by the budget for this activity. If abundant laborers are available, then hand watering may be a viable solution. In some cases, burros have been used to carry large containers of water from the source to the site. Each plant is provided with a small amount of water by a field attendant who walks to each plant with the burro and releases water from the container. Always verify that the water source you use does not have high EC levels. We were involved in a large project at the end of a long dry period, and the pond we were going to use for irrigation had EC levels over twice the value found harmful to plants.

In most temperate areas of North America, irrigation systems are commonly used where extensive container plants are used. Detailed irrigation drawings that indicate appropriate devices in compliance with local codes will ensure a reliable water source. Fixed spray systems are installed for a limited period, usually not more than about three years. During this time, irrigation water is regularly applied, with the goal of encouraging the growth of deep-rooting plants that, within a few years time, will reach the underlying water table, at which time the supplemental irrigation system can be removed from the site.

The design configuration and the pattern you choose may be dictated by some logistical constraints. A project on the Feather River in northern California, for example, needed to plant four hundred acres in one season. The costs in terms of money, labor, and materials for installation, maintenance, and management on a site of this size can be greatly economized by adopting approaches that are similar to agricultural crop production. Many projects on the Colorado River (Anderson and Ohmart 1985) and most recently on the South Fork of the Kern River in central California and the Feather River in northern California have used straight or sinuous rows of plantings, which enable more efficient use of irrigation systems and weed management (Griggs 2009). Aside from initial visual effects, this approach affords several advantages (including cost-efficiency) over the more detailed quadrant patterns that are commonly used for major planting-based projects.

Utility Plans

Utility plans are prepared when utility equipment and facilities are known to exist on your project site. These may become evident during the title search and base mapping conducted earlier. Because of the risk to life represented by high-voltage power lines, high-pressure gas mains, and nearly any utility service, whether it is located above or below ground, it is important to hire a professional to research and prepare this important component of the contract documents. In addition, the liability you face in the event you or your implementation team were to disrupt a utility service, such as a phone line, can run into hundreds of thousands of dollars—as was the case on a site we did on the San Diego River!

Civil engineers obtain mapping from the serving utility company or agency and then plot the locations of each utility facility that occurs on or adjacent to your project site on a utility plan. The utility plan is field verified and typically service companies are hired to spot-check underground locations of pipelines and electrical systems. The findings from the record search and field surveys are then transmitted to the project team and, in turn, are plotted on all related project plans.

What should you do if an existing utility system is in conflict with your restoration project? First, contact the serving agency or company to help you determine the extent of the conflict. In most instances, utilities have preexisting and overriding land rights, which supersede any property rights of your project site. In most cases, you have little recourse other than to redesign your project to work around and preserve the rights and restrictions associated with the utility. In rare instances, utility systems can be relocated, but at the expense of the project sponsor. This is not a simple matter, and professional engineers with experience working with the affected utility should be consulted.

Construction Details

Construction details provide the team responsible for the installation of the project with essential information about how elements of the project are to fit together. Construction details illustrate the relationship of various materials and the way materials are joined and fit together. In many cases, certain elements, such as barrier fences or equestrian gates, require particular materials and construction methods in order to perform as desired. Construction details are engineering drawings consisting of a few views of the same feature—plan view, side view, and cross-sectional view—all provided to give the installation team the information they need to construct the device or system correctly (fig. 9-7).

Plan Preparation

The size and complexity of a project will have a direct bearing on the number of plan sheets and special features presented. Two methods have been used to present plans. On smaller projects, the specifications generally will have been added to the large-format plan sheet. Larger projects

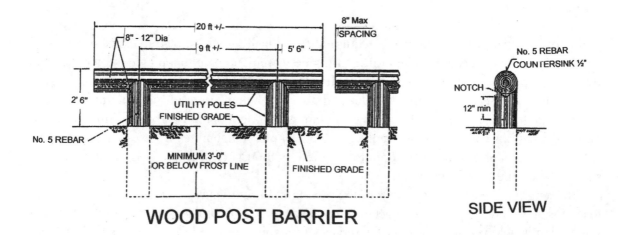

FIGURE 9-7. Construction details for a wood pole barrier, an effective means of controlling vehicular access.

typically have a separate document containing the special specifications and various legal requirements. Together with the plan sheets, the two documents comprise the project plans.

Consider the following guidelines when preparing project plans:

1. Have plans prepared by licensed, registered, and/or certified professionals (e.g., landscape architects, civil engineers, and so forth) in each category of work. Rely on adopted standards wherever possible.
2. Check plan preparation requirements with jurisdictional agencies prior to commencing work; this is especially true for agencies that have permitting authority or jurisdiction over resources being affected.
3. A "Uniform Graphic and Plan Format" should be followed throughout the plan package (e.g., plan format and size, selection and use of symbols and nomenclature, scale, and so forth).
4. Coordinate the development of (a) the plans, (b) the bill of material, and (c) the project specifications to ensure that no conflicts exist among these three documents.
5. In the event of conflicts among the plans, the bill of materials, and the specifications, a hierarchy of precedence must be established that dictates which element has authority.

A suggested format for a project plan follows:

1. Select a plan format during the plan preparation that will be easy to use during the installation phase. We have found that plans prepared on 11- by 17-inch paper can be used during construction and will function very well during installation. While preparing the plan, make sure that symbols and text callouts on the plans are clear, legible, and easily understood in the format you choose.
2. Use consistent terms and references to avoid conflicts with the bill of material and specifications—for example, callouts on the plans should match word-for-word references on the bill of materials and the specifications.
3. Avoid including large blocks of text on the plans, which may create confusion and clutter. The purpose of the plan is to depict the spatial relationships among the various elements that make up your project.
4. Use a system of checks and balances during plan preparation. There should always be an independent review and verification of the plans to ensure correct quantities and to reduce conflicts.
5. Avoid duplicating information—for example, utility information should appear on the utility plan but not on any other plan. Information should be repeated only if it is addressing warnings about sensitive resources or other conditions placed by reviewing authorities.

To assist in developing your plans, appendix 8 provides a list of questions organized into major topics. Reviewing this list will help you address the various aspects of your project and to think things through to a resolution. This list can also function as a springboard to prompt additional questions on your part.

Bill of Materials

In the world of restoration project construction, one of the most important deliverables that the project manager provides to the implementation team is the bill of materials (BOM). At the most elementary level, the BOM is a list of items or parts needed to create the finished project. For the restorationist, the BOM has many functions. It is used as a budgeting tool, an analytical tool, and a communication tool. The collection of these individual pieces of information, and the relationships among them, derived from the restoration plan enables the transformation from idea to actual.

It is in the conceptual stage that the BOM acts as the principal budgeting tool. Establishing an accurate and comprehensive project budget is key if you are going to successfully identify and secure the needed project funding to implement your plan. Building the BOM begins by generating a comprehensive list of all the materials that will be used during the installation phase of the project. Items such as plants are typically grouped according to size and are not identified by species on the BOM. For example, all one-gallon trees are grouped together in the BOM, and a total number of one-gallon trees is identified. Whether you plan to include five *Populus* species, five *Platanus racemosa*, and fifteen *Alnus cordata*—or the same quantities of completely different species—the fact that the plan calls out for all twenty-five trees to be installed as one-gallon plants means that a total of twenty-five one-gallon plants will be tallied on the BOM. This is true for all materials used on the project.

The BOM is a living document. As the design progresses through the development process, the BOM is constantly being updated with the most current project information, forming a running total of the project's estimated cost and the date of a revised page or document. This running total is a key analytical tool used by the project manager to determine whether the project is on budget. Used in this fashion, the BOM becomes a touchstone by which the project design is measured. If the BOM estimate exceeds the project budget, certain project features will have to be revised, or even eliminated, or the project budget will need to be increased to accommodate the design.

Besides its use as a measuring stick for the project budget, the BOM is a communication tool relied on by the implementation team. The BOM communicates the design intent to the team responsible for installing the project by clearly listing every component required for the project. The BOM also functions as a repository for all ancillary item information, such as product data sheets, which include specifications and drawings, manufacturer's recommendations and requirements, and any special notes concerning the use of an item. It also contains vendor information, which will be used by the installation team to help them source each item required for the project.

The BOM plays a vital role in the hierarchy of construction documentation. For projects that are put out for competitive bids, a thorough BOM is invaluable because it clearly identifies what is required in the project and enables all bids to be compared equally. The BOM reflects each and every item contained on the project plans. In addition, the BOM provides a central document that can be used to measure project progress and to help with determining the appropriate progress payments to be made to the installation team.

Following are guidelines for preparing a BOM:

1. List and briefly describe all the items and materials, other expendables, and equipment and tools to be used on the project
2. Include any special instructions related to an item—for example, if sourcing requires a long lead time, this important information should be noted in the BOM and attached as a separate document.
3. Create document control procedures so that only the most current, and thus most accurate, BOM referenced during design development is used for cost estimating and material purchasing.
4. Store the BOM in a central location, accessible to all team members.

Suggested Format for the BOM

The BOM is typically arranged in a tabular format and contains such information as item number, item description, reference information (e.g., drawing or specification number), size, quantity, special notes, unit price, and vendor data. BOMs are typically created and maintained in a spreadsheet program such as Microsoft Excel (fig. 9-8), although more complex projects may warrant the consideration of BOM software, which is developed specifically to manage the wide array of project data on complex projects.

Specifications

Specification is a detailed and precise description of an intended outcome, one that is concerned more with performance and less about how something is done. Specifications focus on functionality and very often include a criterion for measuring performance.

The criteria used in most specifications are well known and are published by government bodies and nonprofit standards development organizations. For example, the American National Standards Institute (http://www.ansi.org/) publishes standards for construction and other industries, and the American Society for Testing and Materials routinely develops standard test methods and specifications covering construction, the environment, and many other subjects. Specifications almost always include some provision that establishes methods and routines for inspecting or examining the performance or adherence to a standard. This is a critical element of the value of the specification: if there is no clear method for evaluation, then there is no clear path to resolve disputes that may arise as a result of differing interpretations of the specifications. Although we cannot eliminate all misinterpretations, good specification writing can certainly reduce them and help avoid costly errors and rework.

Because the typical project requires many specifications, it is critical that these documents be prepared and organized in a consistent, clear, and thorough manner. This is particularly important when contractors and subcontractors are involved. Therefore, we recommend that before you embark on preparing project specifications you first determine who the audience of the specifications will be. If you feel certain that your plan can be fully implemented through the direct efforts

Bill of Materials

Item	Item	Unit of Measure	Estimated Quantity	Unit Price	Item Total
1	CONSTRUCTION AREA SIGNS	LS	LUMP SUM		
2	ROADSIDE CLEARING				
3					
4			53		
5	MULCH	CY	7,980		
6	COMMERCIAL FERTILIZER (SLOW RELEASE)	LB	8,186		
7	COMMERCIAL FERTILIZER (TABLET)	EA	49,280		
8	PLANT (GROUP A)	EA			
9	PLANT (GROUP B)	EA			
10	PLANT (GROUP F)	EA			
11	NATIVE SEEDING	LB	570		
12	PLANT ESTABLISHMENT WORK	LS	LUMP SUM	LUMP SUM	
13	CONTROL CONDUCT	LS	LUMP SUM	LUMP SUM	
14	1" ELECTRIC REMOTE CONTROL VALVE	EA	3		
15	1 1/4" ELECTRIC CONTROL VALVE				
16	1 1/2"				

The "Bill of Materials" identifies all materials components of the project, with the quantity and the item price.

Plant quantities are distinguished according to container size in the Bill of Materials.

If the total Bill of Materials cost is within the project sponsor's budgetary goal, the project can proceed. If not, components are reduced or cut until the cost goal is attained.

FIGURE 9-8. An example of a bill of materials, a summary document indicating descriptions, quantity, and payment type of needed material for a project.

of your core project team, then minimal specifications will likely be needed. At a minimum, a plant summary should be compiled to ensure that all materials will be available (appendix 7). However, if you involve tradespeople and establish contractual relationships for certain tasks to be completed, then not only will specifications be required but also contract provisions. If you anticipate contracting out all, or a portion of, the project work, then involving an attorney who specializes in developing and reviewing contracts may be prudent.

All of your effort on good design and thorough research can be lost to poorly written specifications. It is not a wise course to anticipate that specific individuals will be installing your project or tasks, because unexpected events happen. Successful project managers make effective use of

---

Box 9-1. Acronyms Can Get You into Hot Alphabet Soup!

A state agency developed contract plans for a bridge replacement project. During the environmental process, a small population of an endangered annual plant was found along one side of the bridge approach. The biologist labeled the area as an "environmentally sensitive area." The plan preparer accurately drew the site and labeled it "ESA" because the area was too small to contain the full wording. The project went to construction, and the biologist visited the site to discover that the contractor had parked his equipment in the area. When approached, the contractor said that "ESA" stands for Equipment Storage Area!

Lesson learned: The state instituted a program of spelling out the words for ESA at least one time on each sheet where a designated area is located. Also, the plans are now reviewed by the environmental units prior to advertising to ensure that all mitigation measures are clearly marked or included in the specifications. Mandatory prejob meetings take place with a representative from the environmental department to discuss the environmental issues of the project.

---

standardized specifications and provisions wherever possible. Good projects have failed as a result of poor specifications or assumptions (box 9-1).

Specification Standards

Standardized specifications—such as those published by the Construction Specifications Institute (CSI)—along with CSI's program for specification preparation, organization, and methodology, are great sources for guidance in preparing specifications. CSI's member organization has worked for years with various trades in the construction industry and has developed a standardized methodology for preparing specifications and contracts. Misunderstandings and possible legal battles can be avoided through the use of a specification structure like that developed by CSI (http://www.csinet.org/). The highly variable nature of restoration work necessitates accurate and precise specifications. Clarity is an absolute necessity to ensure complete understanding of what is desired.

A number of commercial vendors provide standard specifications for their products. This will ensure correct application or use of the material. The following guidelines for specification writing will help you through the process:

1. Begin with the end objectives in mind.
2. Describe the expected outcome or product that you hope to attain.
3. Rely on standards as much as possible; do not waste time reinventing something that has been done. Use an existing format.
4. Be concise; do not overdescribe things.
5. Consider whether a drawing or construction detail might better describe what you are looking for.

6. Avoid duplication between plans and specifications.
7. Check your contract language to make sure it verifies the hierarchy of contract documents. Typically, in construction contracts, specifications (special provisions) take precedence over plan drawings.
8. Consider printing the specifications on the plans. There are two different perspectives on this. It is argued that if the specifications are on the plans, then they won't become separated from the plans and the contractor cannot claim ignorance. However, if you have a significant number of specifications, having them printed on the plans can be very cumbersome to review or refer to during construction.
9. Do not use abbreviations or symbols.
10. Do not use the phrase "and/or."
11. Do not use such phrasing as "approval or satisfaction of the contract manager (engineer)" or "acceptable to the engineer."
12. Do not use the phrase "unless otherwise permitted by the contract manager"; permitting is used only when there is a meeting of the minds.
13. Determine how you intend to pay for the construction item.

A suggested format for a specification contains three elements: (1) a summary statement, usually a one-sentence introduction; (2) details of the product using direct and descriptive terms, with exclusive statements rather than inclusive statements; (3) a description of the method of payment for contracting projects and a statement of conditions for how payment will be made.

As-Built Plan Drawings

In many cases, modifications are made to the project plans and specifications at the last minute or in the field during construction (chapter 10). Examples of modifications include minor changes in grading and resoiling on the site; the substitution of one plant species for another because of plant availability; and changes in planting density or planting locations. It is very important to prepare as-built plans as soon after a project has been constructed as possible. The most important plans that need to be updated are the grading plan and the planting plan (including the plant list specification table). These revised plan drawings should be stamped "As-built" and dated. Researchers who may want to evaluate your project sometime in the future will need to review these as-built plans in order to compare observed conditions with a baseline condition (e.g., site improvements or plant installations). In the past, the lack of as-built plans has made it difficult for scientists to evaluate the effectiveness of restoration project work.

Chapter 10

Construction and Installation

Restoration projects are implemented through management practices and techniques and, where needed, construction of the prescribed physical features (improvements) at a project site. Measures (including biota) designed to initiate restoration of the intended ecosystem are then installed. Throughout this process, the work is reviewed for compliance with the plans, permits, and project stakeholder's stated desires (fig. 10-1).

Defining Your Role and Participation

Restoration projects are implemented by a variety of workers with multiple skills and varied experience. These people have such titles as "design-build contractor," "field supervisor," "project supervisor," "independent contractor," "subcontractor," and "site inspector."

Your involvement in the construction and installation of a restoration project will vary depending on a variety of factors, including your training and experience; the capacity of your firm, organization, or agency to perform the work required; policy, regulations, and licensing requirements

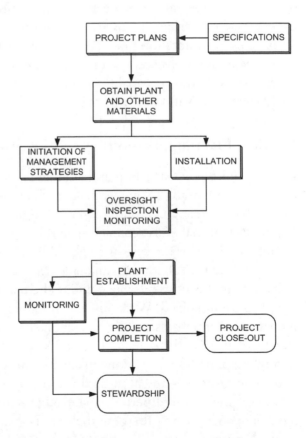

FIGURE 10-1. The implementation phase covers the range from paper plans to in the ground. Although the project may close out at this phase, typically monitoring will continue for several years.

regarding contracting for construction and installation; and the best combined workforce to get the job done effectively and efficiently. You may have prepared the project plans and are now being asked to construct/install the project: this is often referred to as design-build. In this instance, you may have been the project planner/designer and now serve as the "design-build contractor." Or, you may specialize in installing and maintaining restoration projects while working under the direction of a design-build contractor. In that instance, you may function as the "field supervisor."

In another instance, you may have been involved with project planning, design, and plan preparation but now will serve only as the "project supervisor" overseeing the work performed by a contractor who has bid on a set of plans for constructing and installing the project. Or, you may be the "independent contractor" who submitted the winning bid on the plans for construction and installation of the restoration project. You could be the "field supervisor" working for this "independent contractor." Or, you may be a "subcontractor" who has been given the responsibility of constructing or installing only a portion of the improvements at the project site. Finally, it is possible that your only involvement with a certain restoration project is serving as the "site inspector," or one of several site inspectors, making sure that the project is constructed and installed in accordance with project plans and specifications.

It helps for restoration practitioners to understand all aspects of the construction and installation of restoration projects because their roles may differ from project to project. For this reason, practitioners should be capable of functioning as "design-build contractor," "project supervisor," "field supervisor," and "site inspector."

Detailed Planning, Permitting, and Licensing

Planning for the construction and installation of a restoration project is an extension of the project's overall planning (chapters 1 and 3). Preliminary planning for construction and installation should have already been done during the design phase (chapter 6), the determination of the required plant materials (chapter 8), and the preparation of the project plans and specifications (chapter 9). If this is the case, then planning at this stage becomes merely a task of refining, adding more detail to the scheduling, and procuring the required materials and equipment.

Permit applications are made during the planning phase, and permits should be issued when the plans are finalized. Work on-site cannot begin until permits are issued, so, if required, verify the issuance of permits and the conditions that will control the installation. A variety of permits may be required at the federal, state, and local levels in the United States (appendix 9). Ecological restoration practitioners working in other countries will probably need to obtain permits for many of the circumstances and activities listed in appendix 9.

Identifying required permits and making initial applications for these permits should already have happened during the design phase. However, taking the final steps needed to secure the required permits can still be time-consuming. Generally, regulatory agencies that issue permits or other forms of agreements will want to review near final plans (e.g., plans that are 90 percent complete) before signing off on a permit. However, as mentioned previously, permit issuance is more likely if the agency has been involved in reviewing the conceptual plans and early-stage plan drawings (e.g., plans that were 50 percent complete). When permits from multiple agencies are required, the amount of time to work out conflicting differences in the imposed requirements can be considerable.

Typically, certain aspects of construction and installation require the direct involvement of someone who holds a license to perform the work; however, this requirement is not always followed for projects on nonprofit and private lands. Construction requiring the movement of a significant amount of earth using heavy equipment generally requires a grading contractor. The construction of engineered structures (e.g., permanent water control structures) generally requires the involvement of a general contractor, whereas a landscape contractor can construct certain small structures (e.g., pathways, low retaining walls, small observation platforms). The installation and maintenance of container-grown plant materials may require the involvement of a landscape contractor in jurisdictions that require contractor licenses. Installing certain types of irrigation systems also typically requires the involvement of a landscape architect, a landscape contractor, or an irrigation contractor. Applying herbicides to control weeds and to kill invasive plants requires that the individual prescribing and applying the herbicide must be a licensed pesticide/herbicide applicator. Independent contractors who bid on public restoration project work are almost always licensed or registered in the state in which they operate.

Most design-build contractors have someone on staff who holds the appropriate license or registration, typically a licensed civil engineer, a registered landscape architect, or a licensed landscape contractor. Sometimes, conservation or land trust organizations and private landowners perform limited site preparation work and install plant materials and aboveground irrigation systems with the help of farmers using agricultural equipment, which generally avoids the licensing issue. River, stream, and creek restoration projects typically require the involvement of a hydrologist, who is commonly also a licensed civil engineer.

Scheduling and Procurement

For most restoration projects, a myriad of activities need to be coordinated, sometimes in a lock-step manner. In chapter 2, we discussed the creation of a project schedule and the use of Gantt charts displaying the start and finish dates for each activity. At this point in your project, however, you will want to prepare a more detailed schedule of activities.

Factors that can force scheduling changes include the weather, the availability of labor and equipment, lack of funds, and delays in the issuance of permits. Slippage in a schedule can run up against governmental restrictions on the time of the year when certain work can be performed (e.g., grading and erosion control), which can potentially delay project completion. It may take you a long time to procure materials, tools, equipment, and labor. You may have preordered contract-grown plant materials from a native plant nursery. If so, are you prepared if the order is incomplete (box 10-1)? You may also have collected or ordered seeds and other propagules. It may be necessary to establish a temporary nursery for the storage of plant materials prior to installation.

> **Box 10-1. Lesson Learned: Plant Palette Substitutions**
>
> The local office of a statewide agency designed a preliminary restoration plan using appropriate native plant species for the upland portion of the project site. The state office prepared the plans, specifications, and bid package for the project. The agency's state office selected the low-bid landscape contractor for plant acquisition, installation, and maintenance. Having never installed a native plant restoration project, the contractor was unaware that the plant materials needed to be contract grown the year prior to when the site was ready for plant installation. When it came time for plant installation, the contractor could not find a sufficient amount of several of the designated upland plant species at any of the native plant nurseries in the region. The contractor requested permission to substitute cultivars of native plant species and other native plants that were not local to the region. The state office approved their request without conferring with the local office that designed the project.
>
> The agency learned that the local designers of the project need to be included in any revisions to the restoration plan. The agency also learned that sometimes there is sufficient reason to exclude the low-bid contractor when that contractor has no prior experience with native plant restoration projects.

One innovative agency in Washington State, USA, established a Native Plant Salvage Program (King County Department of Natural Resources 2011), with a holding facility to maintain an inventory of native plants typically used in restoration for the region. Volunteers remove plants from sites that are scheduled for construction and later replant this native vegetation at salmonid habitat restoration project sites and water quality improvement projects around the county. This successful native plant salvage program has been in operation for more than fifteen years. There are several factors requiring evaluation when contemplating the salvage of specimen plants (chapter 8). You should already have a list of all of the materials that will be required for the construction and installation of your project (chapter 9).

You will need to arrange for the labor to get the job done. Some regions have nonprofit natural resource employment organizations that can supply trained workers and supervisors. You will need to organize volunteers well in advance of the work and have a contingency plan if they become unavailable. Sometimes, project managers working in rural areas have been able to establish strong ongoing working relationships with farmers and ranchers who have workers experienced in the operation of agricultural equipment that can be used to implement elements of a project.

The demands on time and logistics may be such that contracting some or all of the responsibilities to experienced contractors may be the most efficient way to complete your project.

Equipment and Tools

Selection of the proper tools or equipment needed for the job is essential for a cost-effective restoration project. The wide assortment of equipment and tools available to restoration practitioners

FIGURE 10-2. Small vehicles can be very useful in small areas requiring extensive earth moving. This is removal of sediment caused by illegal grading upstream. Orange County, California. (Photo by John Rieger.)

range in technology (i.e., low-tech versus mechanized equipment) as well as in availability and cost of acquisition or rental. The ease of operation and the level of experience and skill required for equipment operation also vary considerably. In general, it works best to choose the simplest tools or equipment needed to get the job done.

Common equipment used for restoration projects include construction equipment (e.g., bulldozers, backhoes, power augers) (fig. 10-2), agricultural equipment (e.g., tractors, tillers, seed broadcasters, seed drills), forestry equipment (e.g., dibbles, hoedads), landscaping equipment (e.g., shovels, tree spades) (fig. 10-3), and irrigation equipment (e.g., pumps, overhead sprinklers, drip lines).

By working closely with the restoration industry, equipment manufacturers have modified their equipment and designed new equipment to meet the needs of restoration project work (e.g., seed

FIGURE 10-3. A "tree spade" digging a receiving hole for a willow tree. More than two thousand translocated willow and cottonwood trees were placed on this site with a less than 1 percent mortality rate; the trees were salvaged from a construction project upriver. (Photo by John Rieger.)

harvesters, seed drills, sod cutters). Restoration practitioners have conducted studies on the calibration of equipment (St. John et al. 2008) and have evaluated the effectiveness of various types of equipment used in the restoration industry. For example, in chapter 12 of *The Tall Grass Restoration Handbook*, Packard and Mutel (1997) present a good discussion on the use of seed drills for prairie restoration, which is also relevant for other types of grassland restoration.

The desire to have tasks done more easily has led to the invention of brush extractors designed to uproot invasive woody weeds. The Weed Wrench™ (fig. 10-4), available from The Weed Wrench Company in Oregon, and the Pullerbear™ tree and root puller, available from Pullerbear in British Columbia, Canada, are excellent examples of this innovative drive.

Another innovative example, for planting woody cuttings on stream banks, is the "hydrodriller" (Oldham 1989), originally developed in the late 1980s by the staff of the Kings River Conservation

FIGURE 10-4. A Weed Wrench™ is an efficient and easy-to-use tool for uprooting unwanted plants from restoration sites. The clamping foot grips the plant stem, and then a long handle (top of photo) is pulled back using the foot as a fulcrum uprooting the plant. (Photo by Mary F. Platter-Rieger.)

District, in Fresno, California, and subsequently reinvented as the "waterjet stinger" described in detail by the USDA Natural Resource Conservation Service (Hoag et al. 2001). An example of a piece of equipment designed for use in difficult situations is "the "stinger" (Hoag and Ogle 2008), which attaches to the end of a backhoe arm and is used to plant hardwood cuttings in existing riprap.

Kloetzel (2004) provides information on the availability and applicability of a variety of hand and power tools used at ecological restoration project sites. Two relatively new pieces of equipment, designed to mount on the arm of an excavator, include the expandable stinger and the Hy-Gro Tiller™, which are used for performing spot cultivation in preparation for hand planting.

Your choice of the specific tools or equipment to implement your project takes some informed thought. Following are some factors to consider:

- Is there a volunteer workforce capable of performing a lot of work using simple tools but not trained in operating certain pieces of equipment?
- Do you have access to skilled operators of the desired equipment?
- Is the equipment available for rent or for hire?
- Do you know the best equipment and tools to use in certain situations?
- Do you understand the limitations of the equipment?

When contracting work, it is generally best to let the contractor figure out which equipment will best do the job. Contractors often come up with ingenious choices of tools and equipment to perform the required work cost-efficiently. If a contractor suggests employing a piece of equipment with which you are unfamiliar, you may want to ask how the equipment will perform in your situation and to request examples of projects the equipment was used for in the past.

Worker Training

A trained workforce is another underlying factor for successful restoration project work. Training, formal and informal, can take many forms. All employees and volunteers should know their roles and responsibilities and should understand the procedures and techniques for the specific tasks (box 10-2). Those overseeing project construction and installation should be knowledgeable and experienced in all tasks that will be performed by their workforce.

Contractors and subcontractors are required under their license to conduct periodic safety training for all crew members. Safety training should not be overlooked just because licensed contractors are not involved with your project. When working with volunteers or day laborers, it is important to start off each morning with an overview of the important safety measures that must be followed.

When overseeing the work of a contractor or subcontractor, it is the project manager's responsibility to make sure that the contractor or subcontractor is using employees who have been trained in the tasks they will be performing and who understand the safety measures that must be followed. It would be nice if one could assume that this is true; however, experience has shown that it is important to check up on the caliber of the laborers who will be installing your project. A good time to bring up the issue of experience of a contractor's employees is when you are reviewing the project plans and specifications with the contractor, if not already at the contract award selection interview.

Site Inspection

Site inspection during construction and installation is critical to the success of ecological restoration projects (box 10-3). The primary role of the site inspector is to make certain that a restoration project adheres to the restoration project plans and specifications.

The restoration project site inspector performs two functions: a "hard-hat function" and a "soft-hat function." The "hard-hat function" of the site inspector is to observe and sometimes measure

Box 10-2. Restoration Project Highlight: The Use of Volunteers Can Make a Project Cost-Effective and Build Community Support for Similar Projects in the Future

Location: Oyster Reef Restoration in Canaveral National Seashore, Florida, United States

Mortality of oyster reef margins caused by disturbance of the sediment from boat wakes had grown so significant that something needed to be done to save the oyster reefs. Worldwide, oyster reefs have been reduced in area by 85 percent. An innovative oyster reef restoration project in Canaveral National Seashore was developed by The Nature Conservancy, the University of Central Florida, the Brevard Zoo, and partner agencies, organizations, and corporations that relied on a significant number of volunteers to construct and install the oyster reefs.

The plan was fairly straightforward but required the construction by volunteers of almost twenty-six thousand mats, each about 4.5 square feet and altogether totaling almost 2.7 acres. Made of aquaculture-grade plastic mesh, each mat had thirty-six oyster shells attached in an upright orientation. The mats were laid out in a tile-floor pattern on dead oyster bed margins that had been leveled to create a blanket of stable substrate for oyster larvae settlement.

Since 2007, fifty reefs have been restored within Canaveral National Seashore using this method. More than twenty-three thousand volunteers, including many schoolchildren, have assisted this restoration project by not only making the mat assemblies but also placing and securing the mats on the dead reef margins. Monitoring of the reefs has determined that the nearly twenty-six thousand mats have provided for well over 2 million live oysters. An added benefit of the project has been the recent establishment of seagrass at many of the reefs where none existed previously.

Several agencies, organizations, and corporations, along with community involvement, contributed to the success of this project. The labor-intensive mat building process required for this method made the project feasible only by using volunteers. The method developed has been proven successful and can, with experimentation, be applied to other locations and situations. In this case, a large number of volunteers came forward to work on an estuarine restoration project. Such a big number indicates that there is a significant population in the general public interested in participating in restoration projects.

Source: Abstracted from IUCN WCPA Ecological Restoration Taskforce 2012.

the construction work performed by the contractor and subcontractors to ensure that it is in accordance with project plans and specifications. The "soft-hat function" is to ensure that the biotic materials installed on-site comply with project specifications.

Large projects can have more than one site inspector on-site: one experienced with the inspection of grading, electrical, and concrete work, for example, and one experienced with the oversight of restoration project suppliers and restoration project installation contractors.

Box 10-3. Lesson Learned: Site Inspections Are Critical

An agency contracted for the installation of wetland plant species at a stormwater quality improvement project site. The native wetland plants had already been contract grown by a nursery. On the day of plant installation, the plant materials arrived at the project site in the small plastic tublings in which they were grown. The contractor responsible for plant installation picked up day laborers in front of a local market to install the plants. Fortunately, the project designer showed up to find that the laborers were putting the plants into the ground while they were still in their plastic tublings. The laborers were instructed to take the plants out of the tublings prior to putting them in the ground.

The agency learned that it is important to assign a site inspector to the project to oversee the work of the installation contractor.

Restoration project site inspectors are often called on to make necessary adjustments to project plans in the field as situations arise. Sometimes, the contractor makes a suggestion that will make your project more successful, and sometimes contractors just want to cut corners to compensate for underestimations of the time and expense for project implementation. You may be faced with spur-of-the-moment decisions, such as whether you should

- allow a contractor to grade the site to a lesser degree of resolution than what is called for in the project plans.
- allow a contractor to modify the design or placement of in-stream aquatic habitat improvement structures.
- allow a contractor to rip compacted soil to a lesser depth than what is specified in the project plans.
- allow a contractor to deposit a thinner layer of soil than what is called for in the project plans.
- allow a contractor to use erosion control mulch that is not certified as weed-free, as was specified in the project plans.
- allow substitutions of different plant materials than what was specified in the project plans if the contractor cannot locate some of the required plant species.
- allow a contractor to dig shallower planting holes if the ground is harder than expected.

The answers to these questions depend on the specifics of the site and the judgment of experienced individuals. At times, it will be necessary to refer such questions back to the planners for the project because they have the background information of the item being discussed (box 10-4).

A log or journal can be invaluable. Project information to record includes when and where each task was performed, when formal inspections were conducted, the results of those inspections, and documentation, including photos, of any variations to project plans and specifications that were approved. Records of the number of person hours it took a contractor to perform certain

---

Box 10-4. Lesson Learned: Confer with the Designer on Any Design Changes

A geomorphologist/hydrologist designed in-stream fishery habitat improvement structures and biotechnical stream bank stabilization measures for an urban stream restoration project. The plans for the project were approved by a local agency, and the project went to bid. The contractor proposed a modification in the number and placement of the in-channel rock structures. The plans were revised without the knowledge of the lead hydrologist, and the project was constructed during the low-flow period. High flows during the winter following project construction eroded the stream banks and changed the course of the channel, resulting in the loss of adjacent parklands. The contractor blamed the erosion on the project design, and the project designer blamed the damage on the modifications to the design features. The client was caught in the middle of this finger-pointing.

The project lead and the client learned that, while a contractor may make worthwhile suggestions for the revision of stream restoration plans and specifications, it is essential to have the key project designers review and approve of these proposed plan modifications prior to their implementation.

---

tasks will prove helpful in estimating future project work. The log can also be an important information source when producing as-built plans.

Implementation of the Project

Initial activity addresses getting the site ready and then progresses through to the finish of the construction, including documenting exactly what was done.

Removal of Impediments

One of the earliest activities during site preparation is the removal of likely impediments to project success. Impediments that need to be dealt with include invasive plants, nonnative wildlife, and sometimes native wildlife, especially if your site is downwind or downstream (fig. 10-5). Unless prior arrangements have been made, all work must take place within the project limits as identified on the plot plan.

We discuss strategies for the control or eradication of invasive plants and weeds in chapter 12. However, at the beginning of the implementation phase, there are additional procedures available that will not be feasible after construction and installation have begun. For example, you may want to try solarization as a means of killing the soil seed bank. This technique has limitations but can be a simple and effective means of control in the correct circumstances (Katan and DeVay 1991; Lambrecht and D'Amore 2010). You may want to pretreat your site with pre-emergent herbicide to prevent the growth of invasive plants and other weedy plant species. Some practitioners remove and discard the top few inches of topsoil in an attempt to rid the restoration area of the seed bank.

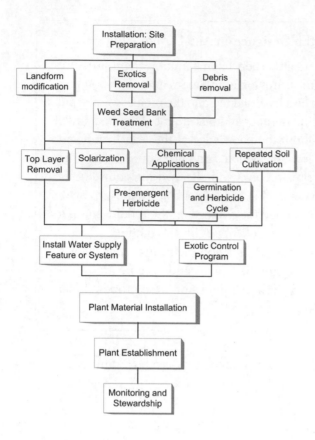

FIGURE 10-5. A flowchart of various site preparation actions prior to installation at a restoration site. Account for these tasks in your schedule to ensure supplies are available when needed.

Other practitioners have tried to force the growth of weedy species by irrigating the site followed by discing in the weedy plants, sometimes more than once. Regardless of your strategy for controlling exotic pest plant species, the best time to start is when implementation begins.

Sometimes, you will know in advance that the presence of certain native or nonnative wildlife species may impede the progress of your project. You might find it necessary to construct exclusionary fencing to limit access by browsing animals. You may need to temporarily reduce the population of gophers on-site or trap and remove beaver, even though the restored habitat may be intended to support beaver in the future. Hunting dogs have been used to chase away geese for the first couple of years in the Upper Truckee River Marsh Restoration (Lake Tahoe Basin, California Tahoe Conservancy 2012b). Brown-headed cowbirds are trapped (fig. 1-4) and removed from locations in southern California where the success of the riparian habitat restoration project depended on the nesting success of the endangered least Bell's vireo. Use of these methods can generate resistance by citizens; be aware of this and build consensus.

Installation of Site Modifications and Infrastructure

Typically, restoration projects require at least some form of site modification and construction of infrastructure in support of the project. It generally follows that the more urban the project location, the more likely the need to modify the site and existing infrastructure.

Where there is existing infrastructure on, above, or below your project site, you may need to take measures to accommodate the feature(s). You may need to remove abandoned structures. If your project is on abandoned agricultural land (old fields) with high groundwater, you may need to remove or break up old tile drains installed years ago. This can be quite expensive and deserves a line item in the budget.

After you have relocated existing infrastructure, you can proceed with any required grading of the project site. This may be a two-step process if the final grading and resoiling requires a finished grade that must reflect a narrow tolerance in elevation (e.g., wetland sites) to comply with specifications. At the same time, you may also be using equipment to reduce soil compaction on the project site.

Sometimes, you will need to get to the other side of a natural area to reach a restoration site with construction equipment. Practitioners in some regions have developed means of installing temporary access roads on top of meadow and wetland sites that, once removed, show very little impact to the original surface (California Tahoe Conservancy 2012a). This can be done by placing soil on top of portable road mats, which are plastic or fabric (e.g., geotextile) laid on top of the vegetation. Also, several types of new low-impact wheeled vehicles, similar to those used in the logging industry, have been invented that minimize soil compaction. Install temporary measures to protect any vegetation that is to remain on the project site and any sensitive areas from accidental damage by equipment. Temporary fencing may also be needed to define and restrict access to planting areas. Sometimes, this "temporary" infrastructure will need to be strong enough to remain in place for several years.

The temporary nursery discussed earlier and in chapter 8 will require fencing and a water supply. The nursery location should be sheltered from the wind and, depending on the species held, may also need protection from the sun. Bare root plants will need to be kept in a dark, cool, moist location or temporarily planted in trenches. Special provisions will need to be made for holding any woody cuttings that cannot be installed right away.

We talked about establishing a water supply and constructing temporary irrigation systems in chapter 7. Now is when you will want to establish a point of connection for your water supply at your project site. Generally, restoration projects do not require electricity at the site; however, a temporary electrical line may be needed to operate pumps or other machinery.

Permanent infrastructure should be constructed at this time. This includes access roads, drainage improvements, water control structures, fencing and barriers restricting access, trails, and visitor viewing areas.

Installation of Biotic Elements

Having healthy plants is essential, but it cannot overcome poor handling and installation. Proper installation of the plant materials (fig. 10-6) cannot be overemphasized.

One responsibility of the site inspector is to inspect the condition of the plant materials that are delivered to the project site. While inspecting the aboveground parts of plants, you should evaluate the vigor (health) of the plants and look for any signs of injury or pests or disease. Inspect the root systems of randomly selected plants grown in containers to confirm root health. Dorner (2002) describes additional issues:

- Is the root ball well rooted in the substrate?
- Are there healthy main roots and plenty of smaller, fibrous roots?
- Are there circling or kinked roots close to the main stem or in the central root zone?
- Are the root tips healthy and actively growing?
- Are there circling roots at the bottom of the container?
- Are the roots mostly at the top portion or on one side of the container?
- Are there roots coming out of the drainage holes at the bottom of the container?

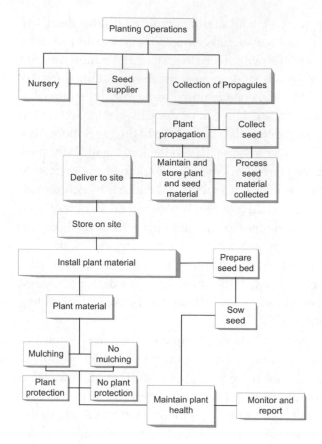

FIGURE 10-6. Planting operations can be involved, so understand the material and tasks needed as well as the time needed from ordering to delivery on-site. Incorporate these time frames into your schedule.

As discussed in chapter 8, the use of the right type of container for growing each plant species combined with periodic inspections of the plants at the nursery where they are being grown should prevent most of these problems from occurring.

Various planting techniques are available depending on the type and size of plant propagules selected. For example, woody cuttings can be inserted by hand, pounded into the ground with a mallet, placed in shallow trenches and covered with soil (willow wattles), installed with a waterjet stinger, shoved through riprap with a stinger, or installed with a hand soil auger, a handheld power auger, or a power auger mounted on a vehicle.

There are proper ways to dig holes for plants and to install container stock. Under certain soil conditions, roots cannot spread into the surrounding areas unless the sides of planting holes are scarified. Plant roots encircling their container need to be pruned. Some plants are unlikely to survive if they are installed too deep or too shallow. There should be no air pockets around the roots, and aerial stems must be erect or nearly so. Volunteers need supervision on these points.

Large blocks of vegetation (sod slabs) can be translocated using a skip loader, a larger sod cutter, or a similar modified tool for scooping up wetland and herbaceous vegetation along with the soil and root systems. Sod slabs have been stacked on top of one another like steps to construct or stabilize stream banks in meadows. If there is a lag time between when the sod slabs are harvested and when installation can occur, the slabs need to be kept in a holding area and watered as needed.

Installation of Plant Protection

Decisions concerning plant protection devices will be resolved during the design phase. Table 10-1 lists several of these devices. The efficacy of many types of plant protection devices that can be purchased or fabricated has been tested with respect to plant survival, growth, and the prevention of browsing impacts (Hall, Pollock, and Hob 2011).

There are numerous causes of plant loss when developing a restoration project, many of which result from the typical physical characteristics of a new site. Generally, new sites are open, with widely spaced plantings that are low in density and diversity, and the plants are small and tender, with roots just starting to become established in the native soil. Placing a container plant into the ground at a size larger than a new seedling bypasses many of the natural processes that cause plant mortality.

When planting in an urban area with a high probability of human traffic, the plants will need protection from trampling. Generally, simple screen devices are sufficient to notify the visitors of new plants that have not grown to a height or stature that makes them readily visible (fig. 10-7). These screens can be enclosed so that they protect the plant from grasshopper and other insect swarms. Although this requires more frequent monitoring, it can save plants that would otherwise be eaten. Plants may be placed in situations that require support until root systems are sufficient to take on one of the important roles of soil stabilization. Planting on steep slopes (figs. 10-8, 10-9, and 10-10) or in windswept locations requires some additional physical assistance until the plant is able to add roots in response to the harsh elements. However, in general, the staking of trees (typically two stakes, one on each side of the tree) is discouraged because the tree trunks may fail to develop sufficient strength to withstand winds or other threats after the staking has been removed.

FIGURE 10-7. A screen may be needed where insect and small mammal damage is expected. This requires monitoring to prevent damage to plants as they grow. If neglected, the plants can be damaged as they grow through the screen. (Photo by John Rieger.)

Table 10-1. Plant Protection Devices

Plant Protector	Description	Trade Names
Seed Protectors		
Plastic Netting	Light (6 mil.) and heavy (12 mil.) netting	
Stretch Netting	Elastic stretch netting (15 mil.)	"C" netting, Tiller net
Wide-Mesh Netting	½ × ½ inch mesh	
Bird Netting	Polyethylene knitted netting; diamond, square, or hexagonal mesh	Avigard® Crop Net
Collar and Screen Direct Seed Plant Shelter	Collar is a one-quart cottage cheese container (bottom removed); window screen attached to collar with plant tie or wire	
Direct Seed Plant Shelter	Two plastic parts quickly assembled in the field; 11 to 24 inches in height; 2.5 inches in diameter.	Blue-X® Direct Seed Plant Shelters
Seedling Protection Devices		
Milk Carton	Cut top and bottom off used milk carton; quart or half gallon size	
Rigid Seedling Protector Tube	Diamond-shaped mesh tube of flexible, UV-inhibited photobiodegradable polyethylene and polypropylene material; 18 to 36 inches by 3.25 to 4 inches in diameter	
Vine Grow Tube	Height 24 to 60 inches ; diameter 3.5 inches; side opening for easy removal and reuse	Plantra® JumpStart Vine Grow Tubes
C-style Tree Tube	Made of recycled polyethylene; comes in heights 12 to 60 inches; about 3.5 inches in diameter; expandable diameter by coupling protectors together	Jump Start™ C-style Treeshelters; Protex® Pro/Gro Solid Tube Tree Protectors
O-style Tree Tube (Tuley Tube)	Preassembled, twin-walled, translucent tube made of UV-stabilized, recycled polyethylene; heights from 12 to 72 inches; diameter ranges from 3.5 to 4.5 inches.	Plantra® Treeshelters Jump Start™ O-style Treeshelters
Tubex® Tree Shelter	4-inch-diameter, twin-walled translucent tube made of UV-stabilized polypropylene copolymer; 24, 36, 48, and 60 inches in height	Tubex® Tree Shelters
Blue-X® Treeshelter	Partially transparent blue-tinted polyester film (PET); 15 to 54 inches in height; 3.5 inches in diameter composed of a single sheet of blue plastic film inserted into a blue-tinted poly sleeve	Blue-X® Treeshelters
Tree Enclosures		
Circular Wire Tree Enclosure	Circular enclosure of steel wire (varying sizes of wire grid) held up by wooden or steel poles	
Game-Deterrent Fencing	7-foot- and 8-foot-high 1-inch mesh; lightweight, UV-stabilized polyethylene; comes in 100-foot rolls	8-foot X-treme Deer Barrier®
Tree Bark Protectors		
Tree Bark Protector	White, opaque, twin-walled corrugated tube that is slit lengthwise and coiled around trunk; 18 to 48 inches in height	Plantra® Tree Bark Protector
Tree Guard	UV-protected white plastic coil	Ross® Tree Gard® Vinyl Tree Guard
Tree Wrap	Waterproof crinkled kraft tree paper	
Tree Trunk Protector	Heavy-duty plastic, 9-inch-tall protector	ArborGuard+®Tree Trunk Protector

Advantages	Disadvantages
Protection from small and large browsing animals.	
Good for deer or elk browse protection.	
	Bird netting is especially lethal to snakes and lizards.
Collar helps to deter gophers. Screen prevents damage from rabbits and insects. Good for seeds and small plants that grow slowly.	Most plants tend to outgrow screen in one season. Requires removal.
One of least expensive methods. Easy installation over plants or planted seed and secure into soil.	Shades seedlings during critical first months of growth. Affords protection for first couple of months. Breaks down after one season.
Reduce animal browsing on tree seedlings.	Requires stakes to hold up tubes. Low branches become entangled in the webbed protector. They generally need to be removed.
Shields young plants from herbicide application. Filters sunlight. Protects plants from mechanical cultivators.	
Traps moisture, raising relative humidity and temperature inside tube. Conserves moisture in arid climates. Protects from animal damage, wind desiccation, rodents, and insects.	Potential for excessive heat inside protector unless tube is vented.
Protects from animal damage, wind, and chemical sprays. Greenhouse environment promotes seedling growth. Vented tree protector allows air circulation.	Shelters generally need to be held up by a stake, in addition to being inserted 2 inches into the ground.
Protects from animal damage, wind, and chemical sprays. Greenhouse environment promotes seedling growth. Splits to allow fast-growing trees to break free.	Recommend using a stake with bird netting to prevent trapping birds.
Protects herbivores, herbicide drift, creates microclimate that accelerates early seedling growth. Produces amplified blue light, increasing beneficial photosynthetically active radiation. Blocks a significant amount of harmful UV light.	Shelters need to be staked. Cannot be reused. May need to be removed from site. Difficult to remove weeds inside protector.
Can vary diameter and height easily. Can protect several closely spaced trees in one enclosure. Browse protection from deer.	Labor intensive to install and remove. Tree limbs can become ensnarled in wire.
Used to fence off a planting area to exclude deer. 8-foot X-treme Deer Barrier® fencing is two times stronger than the regular game-deterrent fencing.	Tall wooden posts need to be installed encircling the planting area.
Shields sapling trees from "buck-rub"/antler damage, rodent damage, and girdling. Protects from mower and trimmer damage.	
Protects tree bark against sun scald, insects, and animals. Protects from mower and trimmer damage.	
Protects tree trunk against rodents, sun scald, severe frost, and windburn.	
Protects base of tree from mower and trimmer damage.	

FIGURE 10-8. Site preparation is mandatory to ensure the wattles can effectively control erosion on the steep slope and prevent the formation of rills and gullies. Naval Base Point Loma, San Diego, California. (Photo by Mary F. Platter-Rieger.)

FIGURE 10-9. A site with freshly installed straw wattles and coconut fiber wattles (darker wattle) for stabilization. Coconut fiber wattles last much longer than straw. This approach was an insurance backup measure in case the initial seeding was not effective. Naval Base Point Loma, San Diego, California. (Photo by Mary F. Platter-Rieger.)

FIGURE 10-10. The same site in figure 10-9 after sixteen months: no significant erosion, wattles in good condition. Sprinkler irrigation was on an as-needed basis for the first twelve months. Naval Base Point Loma, San Diego, California. (Photo by Mary F. Platter-Rieger.)

Plantings in areas where weather conditions are extreme or unpredictable may be best protected by using individual devices. Tree shelters of various types (figs. 10-11 and 10-12) provide protection from wind, herbivory, and water loss. Using individual plant protection devices can be costly and should be evaluated on the likelihood that unprotected plants can survive at some acceptable percentage. Planting more plants and accepting a high mortality rate on the assumption that a specific number will survive despite the elements on-site may be less costly than protecting fewer plants and having a higher survivorship. There are times when no other option is available, however, and establishing a protection device or system is mandatory. Some projects may lend themselves to protecting large groupings of plantings rather than focusing on individual plants. Livestock can be controlled fairly effectively with simple fencing or other barriers (fig. 10-13).

Installation of Erosion Control Measures

Erosion control and stormwater retention are important considerations for many restoration projects, particularly for projects located near streams. Many local and regional governmental agen-

FIGURE 10-11. A tree shelter with three built-in metal rods anchoring the shelter. San Diego, California. (Photo by John Rieger.)

FIGURE 10-12. High deer browsing pressure requires tall tree shelters for plant survival and establishment. Pennypack Preserve, Pennsylvania. (Photo by John Rieger.)

cies have specific guidelines or requirements for how much and how long stormwater can be retained on-site.

A variety of simple erosion control best management practices can be installed right after most construction and installation activities are completed. In the United States, soil conservation advisors from the Natural Resources Conservation Service and local Resource Conservation Districts are willing to assist with erosion control planning. For difficult situations, consider using the services of a Certified Professional in Erosion and Sediment Control.

Erosion control can be particularly difficult on wetland and floodplain restoration project sites. One wetland restoration project in the Lake Tahoe Basin (California Tahoe Conservancy 2012b) used a temporary water-filled cofferdam (Aqua Dam Inc. 2012) to exclude high waters during spring runoff for the first couple of years, while the installed wetland plantings were becoming established. Water-filled cofferdams have been used on numerous other lake, riverine, and wetland restoration project sites.

FIGURE 10-13. Sheep still graze on this slope adjacent to the restoration site. The fencing effectively excludes them from the planted slopes. The bunches of plants on the fence serve as a visual aid to prevent bird strikes. (See box 4-2.) Moffat, Scotland. (Photo by John Rieger.)

Protection against Vandalism

At least four types of vandalism can affect the outcome of restoration projects: random acts of destruction, random accidental destruction, willful acts of destruction, and theft. Strategies for preventing or discouraging vandalism need to be devised in anticipation of the various types of vandalism that could occur at your project site.

RANDOM ACTS OF DESTRUCTION

Typical activities associated with random acts of destruction include pulling up plants, breaking off the tops of sprinklers, pulling out drip emitters, and breaking tree trunks. These events occur randomly and depend on the proximity of the restoration site to a population of individuals with nothing better to do than damage someone else's property.

RANDOM ACCIDENTAL DESTRUCTION

Persons who gain access to a restoration project site for recreational use (e.g., motorcycle riding, horseback riding) can cause significant damage without even knowing it. These individuals are typically former legal or trespass users of the property when the land was vacant. They often unintentionally destroy project infrastructure (e.g., erosion control measures, fencing) and biotic elements (e.g., plantings, seeded areas).

WILLFUL ACTS OF DESTRUCTION

Trespassers who purposefully enter a restoration site to destroy elements of a restoration project (e.g., irrigation systems, monitoring equipment) are typically individuals who do not want a project to succeed. This type of vandalism is the most difficult to control because the perpetrators do not respect signage, fencing, and other barriers.

THEFT

Sometimes individuals steal plants to put in their yards or to use elsewhere for other landscape projects. Stealing of valuable components (e.g., sprinkler heads made of valuable metal) for resale is also a potential problem.

We have found it is possible to minimize vandalism by understanding the social environment within which each project is being constructed. Is the restoration site located in a rural area protected by public land managers or private property owners? Is the project being constructed on open space lands located in the middle of an urban area? Do the adjacent property owners support the project, and are they willing to report vandalism? Does the local community see the project as a benefit to their environment? In general, the greater the degree of stakeholder involvement, the lesser the likelihood for vandalism and the greater the likelihood that vandals will be caught and disciplined.

It is best to do advance planning that incorporates design features to deter vandalism. Following are some measures that can be taken in advance:

- Installing educational signs explaining the purpose of the restoration project
- Installing temporary fencing around the project area
- Hiring a local neighbor to look after the site
- Hiring teenagers to water, weed, and care for plantings
- Installing larger-sized plants in areas where theft is a concern
- Burying the irrigation system in areas where trespassing is anticipated
- Getting neighbors involved in volunteer activities caring for the site
- Conducting public tours of the restoration project
- Educating law enforcement on the importance of protecting the site from unauthorized visitor access
- Hiring security to look after the project site

Many project planners design structures to physically prevent access to a project site (e.g., fencing, off-road vehicle barriers) and inspirational signage (e.g., to psychologically deter vandalism).

FIGURE 10-14. Access control for vehicle and motorcycle while allowing passage of bicycles, horse riders, and hikers. The hanging horizontal log moves away and up when pushed farther, thereby blocking passage. (Photo by John Rieger.)

Off-road vehicles can be successfully controlled using these specially designed devices (fig. 10-14). Some urban restoration project designers have incorporated lighting into associated recreational features, such as along trails; however, one should be aware that lighting can significantly affect the use of the site by wildlife.

In more heavily urban areas, fencing may not be enough to protect the site from all forms of harm. The Mission Trails project had a security guard on duty during the daylight hours; while costly, the project was large enough and the consequences of failure (financial and public works schedule) important enough that it was considered a prudent action. During the year that the security guard controlled access, vandalism was kept to a minimum. We feel that the success of that project is due to this measure, albeit one unusual for a restoration project.

Documentation of As-Built Conditions

After a restoration project has been installed and all construction work has been completed, it is very important to prepare as-built plans (chapter 9). These plans are valuable because sometimes there are significant variations from what was specified on the original plans, for example, in the contouring of the land, the placement of structures, or the layout of installed plants or changes in the plant palette actually installed. Preparing as-built plans is a common practice with governmental agencies; however, project sponsors with limited funding sometimes find it difficult to devote the time necessary to prepare as-built plans. This failure to prepare such plans has made it difficult for researchers to evaluate completed restoration projects for lessons in future project design.

Photo documentation of construction and installation is very important. It is best to establish a set of fixed photo stations. Take photos frequently and regularly at photo stations throughout the life of the project, beginning before any on-site activities have occurred, to demonstrate before and after conditions. It is also important to photograph each of the steps in the process of restoring a site. These before, during, and after photographs will help you tell the story to current and future project stakeholders who may respond to visual material better than to a written report. Photos of volunteers having fun helping to restore a site are a good tool for recruiting volunteers for future projects. We also suggest that you film the work performed on any large projects that may have educational value in the future. Finally, if and when episodic events (e.g., flooding) occur that could affect your project, we encourage you to brave the weather and photograph the event. If the opportunity arises, you may want to arrange to take aerial photographs of the project site. This record may also demonstrate the effectiveness of your restoration project in withstanding or responding to these events.

Installation of Monitoring Infrastructure

In chapter 13, we will discuss the monitoring of restoration projects. Much of your monitoring infrastructure needs to be installed during the implementation phase of the project. Whether they are permanent markers for transects, photo stations, or fixed monitoring locations using GPS coordinates, these monitoring locations should be documented on a set of as-built project plans.

In the past, some practitioners have affixed tags to each individual plant in order to return and evaluate the survival, growth, and condition of each plant. This practice is not as common today because many practitioners measure project success using a variety of monitoring protocols.

Establishment of Maintenance Period

Early in the contracting process, establish that construction bonding will need to cover an extended time period beyond the construction phase to ensure proper maintenance of the plantings. This should be resolved while composing the specifications. However, construction companies are sometimes hesitant to include the maintenance time period within their bond. If this is not feasible, then ensure that a separate maintenance contract is already in place when the installation is complete so there is no interruption of maintenance activities on-site. This is a critical time in the development of your project, and any interruption of maintenance could seriously affect the performance of your project. For example, any interruption in scheduled watering could result in high plant mortality. Also, weed competition at this early stage could retard the normal growth of your plantings or interfere with the germination of your native seed mix.

Project Aftercare

The message in part IV is to keep watch over your project as it develops by executing a well-thought-out maintenance program and conducting monitoring to help assess the progress of the site. Collect the data in a manner that permits analysis. The real acid test of an ecological restoration project plan occurs after it is "in the ground." A site will go through a maturation process, especially in the first few years. Understanding how the site is developing can lead you to new ideas on how to approach design or plant material selection in the future. Young sites are more vulnerable to climatic changes, insect outbreaks, weed invasion, and hydrologic changes. In some cases, the natural processes that you have helped to express will tell you how and when you should decrease, or even cease, your involvement. The only way to know when to stop maintenance or stewardship is to conduct routine monitoring. Depending on the goals of your project, and the uniqueness of the target species, your sampling may range from very simple to quite complex.

Chapter 11

Maintenance and Stewardship

Maintenance involves any short-term activities performed to ensure the development of the project site as intended prior to project completion. Maintenance is generally focused on promoting the survival and growth of installed plant materials during a predetermined plant establishment period to meet established success criteria. Aftercare encompasses both sets of activities occurring in maintenance and stewardship (Clewell and Aronson 2007, 2013; Clewell, Rieger, and Munro 2005).

Stewardship encompasses a wide range of ongoing and, in many cases, long-term management activities intended to promote the maturation of the project site. It typically involves manipulating the processes associated with either the biotic or the abiotic elements of the ecosystem. Stewardship generally follows the plant establishment period but may coincide with, or take the place of, plant maintenance. Our distinction between these two types of activities and their context with the maturation of the site reflects the typical contracting process, in which survival of what is planted is a requirement to ensure delivery of the project in a suitable condition.

Maintenance: Plant Establishment Period

The length of the plant establishment period for restoration projects varies considerably. The plant establishment period is generally based on the amount of time required to ensure the success of the plantings installed at a restoration project site; however, it can be related to any of the success criteria for a restoration project. These success criteria are commonly incorporated into project permit conditions so that the length of the establishment period becomes fixed in the permit conditions.

We have seen restoration projects involving native plant installation with the following defined "plant establishment periods" and associated criteria:

- Survival of all of the plantings for one full year after installation, including replacement of all failed plantings

179

- Survival of a percentage of plantings (or set amount of cover) for one growing season (i.e., one summer in North America) after installation
- Survival of a percentage of plantings (or set amount of cover) for one full year after installation
- Survival of a percentage of plantings (or set amount of cover) for two or more growing seasons (i.e., two summers in North America) after installation
- Survival of a percentage of plantings (or set amount of cover) for two complete growing seasons after irrigation has ceased (assuming that plantings are irrigated for three years, this results in a requirement for a certain agreed-upon level of plant survival for at least five years after plant installation)

We find it is easiest to divide the plant establishment period into three phases: (1) the first year immediately after plant installation, (2) the second year following plant installation until the cessation of irrigation (assuming that some or all of the plantings are irrigated), and (3) the remaining years of the plant establishment period during which there is no irrigation of plantings. Activities associated with the plant establishment period include the following:
- Maintenance of infrastructure (e.g., perimeter fencing, signage)
- Care of individual plantings (e.g., maintenance of plant protection, maintenance of weed protection, weeding around plantings)
- Elimination of unwanted ponding occurring around plantings
- Irrigation of plantings and maintenance of the irrigation system
- Weed management throughout the project site
- The control of exotic invasive plant and animal species
- Any necessary remedial erosion control
- Removal of any debris

It is important to conduct frequent periodic site inspections for maintenance needs throughout the establishment period. A log should be kept documenting all site visits and maintenance activities. This could help in identifying design modifications for future projects or project phases and will also be valuable in estimating the time and materials required for maintenance associated with future restoration projects.

The first full year after project installation is generally the most critical time for a restoration project involving plantings (especially the first growing season). In drier regions, or in areas with little or no rainfall during part of the year, plantings are generally irrigated. Typically during this period dead, dying, or unhealthy plants are replaced and plant mortality is closely monitored. We feel that it is easiest to require 100 percent survival of all individual plants (excluding unrooted woody cuttings) at the end of the first growing season together with the installation of new plants for all failed plantings. Requiring less than 100 percent survival requires a count of plants to establish meeting some criteria that is less (e.g., 80 percent survival). This can be very time consuming.

We also found, prior to implementing this policy, that sites had mortality that just met the needed goal. In many cases, there would be large areas with high mortality. These open areas were quickly invaded by weed species requiring a higher level of labor than previously anticipated. We have also seen situations in which contractors have installed plants just prior to the date of a site review, and thus we ended up relying on poorly established plant materials to meet the success criteria.

Sometimes, plantings receiving supplemental water are irrigated for only one growing season. In drier areas, and in areas of seasonal drought, plantings are commonly irrigated for an additional one to two years. During this time, the amount of water that plantings receive and the frequency of watering are generally gradually reduced. The decision as to when to stop watering altogether should rest in the hands of the site manager, although it may be fixed in project planning documents or permits. Generally, requirements for plant survival are adjusted downward from the 100 percent level required at the end of the first growing season.

In many cases, regulatory agencies require a period of one to two years when there has been no irrigation. Such a period of continued plant survival and growth is often required in permits, especially if the restoration project is providing compensatory mitigation for habitat loss. Although there is no application of supplemental water during this period, other maintenance activities are likely to continue (e.g., weeding around plantings and weed management). Typically, plantings are monitored during this period to ascertain how they are doing without supplemental water. Some plant mortality should be expected, and any concern is relevant only in terms of whether the plantings are on a trajectory to achieving the mandated success criteria. At the end of the predetermined plant establishment period, the plantings are assessed to determine whether they meet the criteria for plant survival and sometimes growth and cover.

Stewardship: Restoring Ecological Functions

The goal of stewardship is to address the whole site, including those ecological processes needed to sustain the desired vegetation community. The intent is to nurture those habitat elements requiring longer time frames to develop and establish a positive trajectory in the overall maturation of the site. This would include deliberate actions focused on creating or providing a specific element of a species' required habitat. Stewardship involves long-term management, which ensures the long-term viability of a project site from controllable damage, such as vandalism and other human-based disturbances. Urban stream projects are excellent opportunities for this type of stewardship (box 11-1).

Stewardship may also include continued weeding and plant care. The concern is less for the survival or functioning of introduced independent elements, such as plants, berms, and streams, and more for the continuation of restored functions and values.

As briefly discussed in chapter 1, two strategies exist for restoration projects: construction and management. In this section, we discuss the reestablishment of ecological processes through various management activities or by simulating what a natural process might do on a site.

Stewardship is involved with making conscious modifications to the vegetation, water regime, or substrate to promote the occurrence or increase of specific or targeted species. This can be accomplished in multiple ways. Typically, it involves the return of a process that has been suppressed

> **Box 11-1. Creative Solution: Encouraging Public Involvement**
>
> A community organization worked with city planners and engineers to daylight a section of an urban stream that had been buried in a pipe for half a century. Following excavation and removal of the pipe, native plants were installed along the protected stream bank of the "restored" stream channel. Because this project was located in the middle of an urban population, there was concern that the plantings could be damaged or removed by young, naive people. The city hired local teenagers to assist in the restoration efforts by watering and monitoring the progress of the plants. These "citizen restorationists" looked after the plants and educated their peers as to the importance of this project for beautifying their community. Vandalism was not a problem at this project site.
>
> The project manager and implementing agency learned that recruiting a local workforce for site maintenance can help educate the public about restoration projects and can limit vandalism.

or controlled into atypical patterns by human intervention. Among the most common of these processes is fire. Restoration of the prairies of the North American plains and management of the conifer forests and scrublands of the southwestern and southeastern United States both involve the use of fire in addition to standard methods of vegetation management. In Great Britain, controlled burning or swailing is commonly used in managing the heathlands, in addition to grazing and mowing.

Fire suppression is one of the most notable interruptions of a natural process in the world's many pyrogenic ecosystems. Returning fire into a system will alter the fuel load of a site, change microhabitats, and promote changes in vegetation types. However, fire in the urban context would normally not be feasible for various societal and economic reasons. What can substitute for fire, and how can other actions replace the effects that fire would normally create? For some ecosystems, mowing or grazing will approximate the general physical effect of fire, with the exception of the nutrient cycling; however, studies have shown that burning is more effective in restoring desired conditions (Weekley et al. 2011). Using a surrogate management strategy may not completely replace all of the effects obtained by the original function.

In some ecosystems, the most prevalent cause of change is the occurrence or nonoccurrence of waters: flooding or long-lasting drought (box 11-2). Many systems that once relied on flooding have become isolated from the typical flooding regime due to the construction of dams and levees and because of channel realignment and channel incision. Sometimes, it is possible to simulate flooding through releasing water from reservoirs or diverting from existing watercourses no longer capable of flooding the site.

Historically, many of the landmasses had large herds of grazing animals. Not only did these grazing animals change the physical state of the vegetation, but they also trampled the soil with their sharp hooves and promoted the vertical redistribution of soil nutrients (fig. 11-1). Grazing is a significant strategy in maintaining the anthropogenic heathlands of Great Britain as well as the

Box 11-2. Lesson Learned: An Episodic Natural Event Can Be a Good Thing

A riparian corridor restoration project was installed on a floodplain a considerable distance from a creek. To ensure the survival of the riparian plantings, an extensive irrigation system was installed to bring water to each plant installed. Although the plantings would be watered only during the dry summer and early fall for the first three years, the expense was justified because failure was not an option. The following winter, the floodplain was inundated for several days—enough to saturate the entire soil profile. The flood also wiped out all of the irrigation lines, large and small.

Faced with a difficult decision as to whether to replace the irrigation system, the project manager decided to wait and see how the plants did without irrigation in the early part of the next summer. By early summer, the plants still looked healthy and continued to thrive the rest of that summer and fall. Excavating a few root systems allowed the project manager to determine that the root systems had grown deep enough to be in contact with the groundwater table. The decision was made to not replace the irrigation system and save money.

The project manager learned to adapt to current conditions not anticipated. By taking a wait-and-see approach, in some cases it may be possible to prevent cost overruns caused by natural events.

FIGURE 11-1. A Konik horse, a small primitive horse from Poland, is used here as a means to control biomass in one of the marsh tracts at Hickling Nature Centre marshlands, in England. (Photo by John Rieger.)

restored hay meadows of western Europe. Methods have been developed using hoofed animals to prevent or control desertification. The land can benefit from using a "holistic approach" (Savory 1998) to managing livestock. Today, many sites cannot support even a few grazing animals, let alone a herd. However, mowing may be an acceptable substitute. In some cases, the equipment is modified to churn or disrupt the topsoil layer, as in restored European hay meadows. These examples represent cultural or semicultural restoration targets, as discussed in Clewell and Aronson (2013).

In the Scottish Highlands, recreation is given priority over biological goals: the countryside is managed primarily for deer and grouse. As a result, two factors—the burning for heath and high populations of deer—have retarded the establishment of new forests. A deer exclosure fence is the primary approach for reestablishing and expanding the severely limited Scots pine forest (fig. 11-2). The seeds of several species occurring in the adjacent "old growth" forests are allowed to germinate and to avoid being eaten by deer, the primary stressor on forest restoration in northern Scotland. The process takes several decades because after even one decade trees are still suitable forage for deer. A number of nonprofit organizations have fenced extensive areas, working in cooperation with the forest agency and a few enlightened landowners.

FIGURE 11-2. In this long-term project in northern Scotland by the volunteer-based Trees for Life conservation charity, a large deer exclosure adjacent to a remnant Scots pine forest allows for natural recruitment. In places, more than twenty seedlings per square yard were observed. Cannich, Scotland. (Photo by John Rieger.)

FIGURE 11-3. A boardwalk in this coastal dune complex has controlled wandering over the dunes, allowing vegetation to reestablish. Carmel, California. (Photo by John Rieger.)

Fencing and boardwalks function to control or influence where and how humans access a site. Several creative solutions have been devised for widely different ecosystems (fig. 11-3). Depending on the habitat, some pathways may provide the only access into the site. Fencing is usually not needed in these situations. It may be necessary to control access to vehicles while permitting pedestrian access using various gate and pole arrangements.

An interesting situation occurred on one of our sites where we were developing habitat for an endangered bird species. The site is approximately fifty acres and planted predominantly with willows (*Salix* spp.) and a few other appropriate riparian species. The site is adjacent to a mature riparian forest area. The region had undergone a period of five years with subnormal rainfall, and all of the naturally occurring riparian systems were stressed. The trees' reaction to this stress was to drop limbs, which limited the amount of leafing. This resulted in a more open canopy, a condition favorable to the insect pest known as the flat-head borer beetle, which exists in low numbers in normal systems but tends to increase in number when the canopy opens. The newly planted site was extremely attractive to the beetle, and we discovered that many of the willows had become infected. An entomologist consulted with us on this problem. We learned that there is no known insecticide to control the pest. The only thing that limits the insect is frost-temperature weather. That would not happen, if at all, until winter, some six months away.

However, upon inspection, we discovered that the willows were not being killed but were only losing the main stem and sending up several sucker branches. The effect was to make a bushier plant that promoted the specific type of nesting habitat that we desired. In contrast, healthy or less stressed plants were able to prevent infestation and did not change their growth form. We learned that the interactions of insect and plants can sometimes lead to positive outcomes.

Depending on the species involved, habitat management can be an annual activity or one that cycles every few years. The replacement of some events is necessary to simulate the occurrence on-site under significantly different circumstances. Having the various habitat elements present on-site for a species is needed for residency.

Restoration of Cultural Ecosystems

It has long been known that the many indigenous peoples of the world actively managed their lands for the benefits they could derive from the flora and fauna. Periodic burning of the undergrowth was a routine followed by tribes in California and other areas of the western United States. This maintained areas in ways that promoted the presence of favored animals used for food, utensils, and clothing (Anderson 2005). Marshes of many types are manipulated to promote specific plants used in basketry (Rea 1983), promote fish for food, and provide numerous other materials for living (Blackburn and Anderson 1993).

In many regions of the world, restorationists have been working with indigenous peoples to restore cultural ecosystems so that native peoples can continue to gather resources from these areas in addition to continuing their traditional spiritual practices. This relationship of people to the land has gained attention with the recent efforts to restore the cultural and biological environment of the Marsh Arabs of Iraq (box 11-3).

Box 11-3. Restoration Project Highlight: Restoring Water Flow Is Commonly the Key Ingredient in the Restoration of Large Wetland Areas

Location: Restoring the marshlands of Iraq in the Tigris-Euphrates river system

In the late twentieth century, the regime in power in Iraq drained the marshlands in southern Iraq as a means of controlling the rebellious Shiite Muslims, who used the marsh to hide. This action also seriously affected a group known as Marsh Arabs, or 'ma'adan', who had occupied this area for millennia (Alwash 2013). A study released in 2009 showed that the marshlands were reduced by 90 percent. The Marsh Arabs depended on resources obtained from the marsh for their entire livelihood. Even though this major impact to the marsh was fairly recent, it has had a major lasting impact.

The marsh is at the confluence of the Tigris and Euphrates Rivers. The marsh formed shallow lakes, mudflats, and deepwater lakes, all providing important habitat for a wide variety of resources used by the Marsh Arabs. At present, water diversion is a major deterrent to marsh restoration. The demand for water by agriculture and oil exploration has reduced water flow into the marshlands significantly, and that demand increased during the first Gulf War. Despite wide international interest in restoring the marshlands, there remain significant issues related to water (Alwash 2013). The Euphrates has dams in Turkey and Syria, and the volume of water released is controlled by those countries' needs, resulting in only one third of the necessary quantity entering the marsh. Creative hydraulic projects on the Euphrates have helped to restore approximately 45 percent of the marsh in that area, with limited populations of birds, fish, and other wildlife returning.

Water management and planning are definitely the major factors affecting the overall continued persistence and restoration of the marsh. Efforts have progressed toward international agreements. Development of plans for a restoration program has identified the need to establish a research field laboratory, to bring the community into the process by education, and to create a database to establish a baseline. Coordination among the various governmental bodies, both local and national, has not been good, and progress on these initiatives is marginal. On a local level, serious disputes between tribes persist. Turkey has built more than twenty-two dams, with more planned. Water diversion on the Tigris River has significantly reduced water flows by one third (Hammer 2006). National and regional planning has failed to provide the overall vision for sustainable development of the marshes. Still undetermined is a plan that accounts for water needs among its primary users. Creative solutions are being proposed to provide the pulsing freshwater desperately needed to return at least a portion of the original marsh back to its historical function.

In lieu of international and national progress, local projects seem to be experiencing the best results. A product of discussions among all members of the community has produced a number of priority actions. These include securing funds by approaching oil companies to contribute as a social responsibility; providing support for the community to develop a restoration plan with a monitoring requirement; incorporating the traditional practices of the region into the restoration process; documenting the oral history of the region as a means of providing a framework to develop restoration projects and benefiting from current restored areas so that other areas may be restored.

Many lessons have been learned from the Iraq marshlands situation. The first is to have open communication and agreement among all potential parties involved. International and national interests, and corporations as well as local communities, must all have a voice. There must be legislation to protect the areas undergoing restoration as well as the creation of a program that provides funds to locals so that they may initiate projects at that level. The government needs to fund these "grassroots" efforts and to support the local initiative. Last, in connection with these efforts, the need exists to start small and build up as the learning process begins and skills are acquired.

The future of the marshes can succeed only with the inclusion of the Marsh Arabs and with sufficient water to support the ecological functions of the marshlands (IUCN WCPA Ecological Restoration Taskforce 2012).

Chapter 12

Weed Management and Invasive Species Control

Weed management and the control of invasive plant species are critical first-year activities and generally must be carried out for several more years until the desired vegetation becomes well established. In fact, many professional practitioners consider weed and invasives control to be the single most important activity on a restoration project site after plant installation. Invasive species control is sometimes the most time-consuming, labor-intensive, and costly component of an ecological restoration project.

Defining Weeds and Invasive Species

A *weed* in agronomic terms is any undesired, uncultivated plant that grows in profusion so as to crowd out a desired crop—in our case, desired native vegetation. Weeds are typically nonnative plants that are invasive—that is, they aggressively compete with the native vegetation. From an agricultural point of view, many native plants can be weeds if they interfere with the growing or harvesting of agricultural crops. From an ecological point of view, many ornamental landscape plants can become weeds. From the point of view of a restoration practitioner, weeds are invasive, generally nonnative plants that interfere with the development of a restoration project site.

Invasive plants are not always nonnative species. The forests of the southeastern United States demonstrate this in places where lowland forest species were not cleared for agriculture because the land was not suitable. After agricultural activity was abandoned, the various species of the lowland forest were able to invade the large expanses of fallow field. With no seed bank and few numbers of the upland forest species represented, the lowland forest was able to expand its range and take over an area that was previously a different forest assemblage (Clewell and Aronson 2013).

An "invasive species" is any species (usually nonnative) that occupies space and uses resources in a landscape that would normally be occupied by native species. When we use the term *invasive species*, we are usually talking about nonnative invasive plant species; however, there are also many invasive animal species, including aquatic organisms, that can adversely affect the outcome of a restoration project. Nonnative invasive plant species are generally the most aggressive "weeds" on, or adjacent to, a restoration project site. This is because many of these invasive plant species evolved in a different region of the world and thus have a competitive advantage because they are no longer controlled by their natural predators. The term *control* here means eradicating, suppressing, reducing, or managing invasive species populations, preventing the spread of invasive species from areas where they are present, and taking steps, such as the restoration of native species and habitats, to reduce the effects of invasive species and to prevent further invasions (US Federal Register, February 8, 1999, Executive Order 13112).

We use the terms *weeds* and *invasive species* interchangeably. Other terms commonly used for weeds and invasive species are *alien species, exotic species, exotic pest plant species, invasive exotics, exotic invasives, noxious weeds,* and *wildland weeds.*

Need for Weed Management

Depending on the type of vegetation or system in which you're working, you may find it acceptable to allow certain weeds to become established on your site. However, in most cases, it is important to control the weeds because they can seriously affect the development of your site. One of our projects had a three-month interruption in plant maintenance following plant installation. This hiatus was sufficient to allow a lush growth of weeds to establish. Several thousand dollars later, we were able to clear the site of many weeds and found that all of the native plants were smaller and less well developed than on parts of the site that had not been subjected to the lush weedy growth. In this example, weeds clearly retarded the growth rates of the installed plants, and the density of the weeds directly affected the amount of growth by those desired plants. This effect may have consequences later if you have planned and budgeted limited resources for maintenance.

Of course, the weed species involved are very important in your maintenance program protocols. Some weeds are annuals and produce vast quantities of viable seeds, whereas other weed species may be woody shrubs that take longer to propagate. There are numerous weed species that do not overpopulate an area and cause negative effects and therefore do not require active control. We consider these species to be low priority based on their characteristics in relation to disruption of restoration sites.

The key to any weeding effort is a consistent program routinely executed. It does little to weed aggressively for a short time and then disappear altogether for a year before re-initiating a weeding program. Attacking weeds only once with herbicides or hand weeding may actually worsen the situation (Murphy et al. 2007). If your project relies on professional workers to perform these tasks, then you should have sufficient funds budgeted to get the site through at least the first growing season and preferably at least two or three seasons.

Most restoration sites are degraded areas where weeds typically flourish, preventing the establishment of native plants. Some invasive plants change ecosystem processes, such as hydrology, soil chemistry, and fire regimes. Many of the weeds that have displaced native groundcovers are prickly or spiny to the touch, making maintenance in restored sites challenging. Tall weeds can overtop and shade new plantings, causing reduced plant growth and vigor. Weeds also compete

---

Box 12-1. Lesson Learned: Using an Inexperienced Crew Can Send You Back to Square One

An agency spent considerable time and money to develop a native plant revegetation plan for upland areas near a creek. Additional expense was associated with the acquisition and installation of the appropriate native plants and irrigation system. As is often the case, nonnative weeds grew up quickly throughout the areas between the plantings. The maintenance crew responsible for maintaining the plantings did not know the difference between native and nonnative plants. Because the weeds had grown quite high by the time they got around to maintaining this site, the crew ended up cutting down a significant number of the smaller native plantings. Subsequently, the agency had to purchase and install additional plants the following year, adding to the expense of the project.

The agency learned that it is important to either hire experienced contractors familiar with restoration project site maintenance or train in-house personnel on how to tell the difference between native and nonnative plants.

---

with native vegetation for available soil moisture. Root systems of some weeds can outcompete native species even though the aboveground plant may not dominate the space.

One very important aspect of weed management is the training of maintenance crews to distinguish between the desirable and undesirable plants on a project site (box 12-1). Maintenance crews should be given field instruction in recognizing the native plants that have been installed at the project site, and they should also be able to recognize the major weed species that need to be controlled. Some project managers have had to replace a large number of installed plants that were cut down or inadvertently trampled by untrained workers. Some restoration practitioners create exhibits of the "good" and "bad" plants by collecting live samples of each species, while others go to the added effort of preparing booklets showing the weedy species. One easy way to do this is by placing live specimens on top of a photocopier.

Developing a Weed Management Program

Most ecological restoration practitioners will need to develop their own weed management program tailored to the needs of their restoration project site. Refer to box 12-2 for sources of information for controlling weeds and invasive plants.

Weed control at ecological restoration project sites requires different strategies, techniques, and schedules than those used in agricultural and landscaped areas. Weeding in the latter environments requires repeated mechanical, chemical, or small-tool destruction of all weeds. Typically, weed control in agricultural and landscaped areas is performed several times per year, every year, whereas the timing of weed control at restoration project sites depends on the weed species involved, the extent of the infestation, and the degree to which the weeds may interfere with the successful establishment of desired native plants.

Box 12-2. Helpful Websites for Invasive Plant Management

European Weed Research Society
http://www.ewrs.org

Introduced, Invasive, and Noxious Plants, Natural Resources Conservation Service,
United States Department of Agriculture
http://plants.usda.gov/java

The National Invasive Species Counsel
http://www.invasivespecies.gov

Invasive Plant Atlas of the United States
http://www.invasiveplantatlas.org/index.html

Weeds Gone Wild: Alien Plant Invaders of Natural Areas (Alien Plant Working Group)
http://www.nps.gov/plants/alien

Because the goals and objectives of restoration projects differ from agricultural and landscaped areas, weed control can be more specific—for example, targeting only specific invasive nonnative weeds, and not all weeds, on a site. Generally, after controlling weeds for two to three years, the frequency of control efforts can be significantly reduced if not discontinued.

Only the most invasive and disruptive weeds require control, thus it is sometimes possible to accept or even encourage the presence of certain "weeds" that may have beneficial values, such as the control of soil erosion. This is similar to the agricultural use of such weeds as wild mustard between rows of trees in orchards.

In some instances, restoration practitioners use the technique of installing a nurse crop. A nurse crop is not necessarily a species, native or exotic, that is appropriate to the ultimate goal of the project, but it is an intermediate step on the way. This technique is used when faced with some type of challenge that native plants are unable to overcome in a short time period. For example, by preempting available space for colonization, a nurse crop can provide immediate cover to discourage other exotic "weeds" from establishing in large numbers. Nurse crops also stabilize soil; add nitrogen, if needed, by planting annual legumes; add organic matter to the soil as the nurse crop dies; and provide temporary shade or windbreak or some other needed function quicker than native species. It is important that you know the relationship of the nurse crop to the target species you are trying to promote. In some situations, exotic species can get out of control and outcompete the targeted species. This is a serious consideration you will need to evaluate prior to adopting it as a technique for use on your site.

One common goal of weed management programs for restoration project sites is to create a mostly native groundcover that will resist invasion by nonnative weeds. This goal may be reached by forcing the sprouting of weed seeds, which are then tilled into the soil or otherwise killed by herbicide or fire. This should eliminate or significantly reduce the weed seed bank in the soil. Reducing the soil seed bank is followed by seeding the site with desirable native species, typically forbs and grasses, before or shortly after installing woody native plants.

Weed Control Principles

The following weed control principles apply to restoration sites and may also apply to natural areas and wildlife management areas:

1. Survey your project site frequently to locate and map any new infestations of weeds and invasive plants. This is assuming that you have identified and evaluated weeds and invasive plants during your site analysis (chapter 4) and have attempted to control or eradicate weeds and invasive plants of concern during site preparation (chapter 10).
2. Aim weed control activities to very-high- and high-priority, targeted weed species.
3. Use selective control techniques on the targeted weeds. Avoid using large-scale nonselective techniques, such as discing, except when preparing a site to be seeded.
4. Control weeds before they have set seed that year. If the weed has already flowered but the seeds have not yet been dispersed, cut off the seed heads and bag and remove them from the site, or remove the entire plant from the site.
5. Weed control is best done as soon as feasible after identifying the infestation. This is particularly important for the highest-priority weeds, which can rapidly expand their range.
6. Control of weeds should be based on the plant's life cycle. Therefore, use methods appropriate to whether the plant is an annual, a biennial, or a perennial. You should follow up on the success of any approach regardless of the life cycle of the species.
7. If the weed has not set seed and will not regenerate itself from the root system after pulling, uprooting, or hoeing, place the weed on the ground where it was removed, to mulch that spot.
8. Uprooting, pulling, or hoeing weeds is best done when the soil is moist.
9. Limit the use of chemical herbicides to only those perennial plants that reproduce vegetatively and are not easily controlled with other methods. Chemical herbicides may pose threats to restoration areas through toxicity to mycorrhizal fungi and soil microflora.
10. Avoid disturbing native groundcovers while controlling weeds.
11. Encourage native groundcovers, especially those that reproduce vegetatively, by removing weeds next to existing colonies.
12. Reseed or replant large bare areas with appropriate native species so that weeds cannot easily re-invade the site.
13. Keep a minimum of a three-foot-diameter area around installed trees, shrubs, and groundcovers free of weeds.
14. Use locally collected or locally appropriate native seed mixes for cover crops. Avoid the use of seed mixes with exotic species. Be aware that there are nonnatives that do not spread, so consult with local experts for your area. In many cases, the behavior of nonnatives is not fully understood, and you may inadvertently introduce a problem onto your site.

Developing a Priority Rating System for Weeds on Your Restoration Site

It is important to identify all of the weeds on your project site so that the most aggressive weeds receive focused control. Weeds deemed nonthreatening or temporarily beneficial are left undisturbed or are controlled only after the high-priority weeds have been controlled. This is because (a) a large number of nonnative weeds normally exist, (b) it is often impossible to control all of the nonnative weeds in restoration areas, and (c) a wide variety of wildlife species can take advantage of some attributes (e.g., seed production) of weeds.

The following priority rating scale was developed to assist restoration project site managers with preparing a site-specific weed control strategy. After identifying the weeds on your project site and on adjacent properties (the latter is especially important for weeds that spread by windborne seeds) and researching their ecology, place each species in one of the following four categories.

PRIORITY 1: VERY HIGH PRIORITY

Priority 1 species are perennial weeds that are generally the most aggressive weeds on restoration sites. With a few exceptions, they reproduce mainly vegetatively (underground rhizomes in most cases) and secondarily by seed. Vegetative reproduction permits them to spread rapidly and exclude other groundcovers. All Priority 1 species are seen as a threat to the integrity of restoration areas and should therefore be eliminated from the entire project area.

PRIORITY 2: HIGH PRIORITY

Priority 2 species are mostly biennial. They are very aggressive and commonly will create solid stands. Priority 2 weeds are large and often spiny, making them a threat to newly planted restoration areas and making restoration site maintenance difficult. Priority 2 species should be controlled for the first three years after installing groundcovers by seed in restoration areas, with an attempt made to eliminate them from these areas.

PRIORITY 3: MEDIUM PRIORITY

Priority 3 species may be annual, biennial, or perennial. They are relatively low growing and generally not disruptive to restoration plantings or to maintenance of restoration project sites. Normally, Priority 3 weeds need not be actively controlled. However, in certain areas where a specific concern exists, control of Priority 3 species may be warranted.

PRIORITY 4: LOW PRIORITY

Priority 4 weeds may be annual, biennial, or perennial. They are low growing, are not disruptive to restoration plantings, and do not interfere with restoration site maintenance. Priority 4 species do not require active control.

Weed Control Strategies in Degraded Natural Areas

A creative and effective approach to weed control was developed in Australia by the Bradley sisters, who were working on bush regeneration (Bradley 2002). This sequence of steps is even suited for one person doing hand weeding. However, if you have several people available, you can divide up your site to take advantage of the increased labor. The first underlying principle is to always begin weed control in relatively undisturbed natural areas and then gradually clear toward the more heavily invaded areas, allowing time for native vegetation to move into cleared areas. The second principle is to try to keep from disturbing the soil any more than necessary while removing the invasive plants, because undisturbed native soil with mulch resists the establishment of invasive plant seeds. This approach is useful for areas with limited budgets, motivated volunteers, and plenty of time to accomplish established goals. Applying the following sequence of steps is best for the eradication of weeds in, and adjacent to, natural areas being encroached upon by invasive plants, but it can also be used on restoration sites.

WEED ERADICATION: BRADLEY METHOD

1. Prevent deterioration of areas not experiencing weed infestation. Start by getting rid of weeds that occur singly or in groups of four or five in areas where native plants are dominant. Check once or twice a year for missed weeds. Depending on the species of weeds and their ecophysiological characteristics, checking may need to be more frequent.

2. Improve the next best. Choose a place that you can visit easily and often, where the native vegetation is pushing against a mixture of weeds and natives, preferably not worse than one weed to two natives. Start with a strip about twelve feet wide and no longer than what you can cover about once a month during the growing season. Give the natives time to move into the weeded areas. If this boundary is on a steep slope that might erode, clear a number of patches instead, but still no more than twelve feet from the vigorous native vegetation. Let a few months go by before you lengthen the strip. Your experience will dictate whether to make the strip longer or shorter.

3. Hold the advantage gained. Resist the temptation to push deeper into the weeds before the regenerating natives have stabilized each cleared area. The natives need not be very tall but should form a dense groundcover. The Bradleys recommend excluding light from the ground—a very important factor because weed seedlings consistently appear in bare soil at the edges of paths and clearings, even when relatively undisturbed and surrounded by dense native vegetation.

4. Cautiously move into the extensively infested areas. When the new growth consists almost entirely of native species with only a few weeds, it is safe to move farther into the weeds. Do not start to clear a block of solid weeds until you have brought the good native vegetation right up to that area. Solid infestations of weeds can be worked on at the edges by forming peninsulas of weeds, small clearings less than six feet in diameter. Also, conduct spot weeding: removing a single large weed plant next to a native plant enables it to grow faster. There is no reason to hurry this process; much more is gained by allowing the native plant to establish well before removing other adjacent weeds.

5. Keep accurate records. Make periodic surveys, and map the weed infestations. Mapping is useful to show stakeholders and governmental agency personnel the progress of the work, and for later reference in planning future work.

WEED ERADICATION: SATELLITE COLONIES

We emphasize the Bradleys' recommendation of focusing on the isolated satellite colony. The reason for this is the phenomenal rate of area increase by these small satellites. The rate and combined areas of increase typically exceed the single core area. Figure 12-1 illustrates the overall effect of satellite colony increases compared to the area increase exhibited by the large core area. The figure assumes the area of the single core colony is equal to the fifty colonies. For purposes of this example, assume each colony, regardless of size, increases outward only one foot each year, an increase in the diameter of two feet. The figure dramatically shows how the fifty colonies can overtake an area more effectively than the single core population.

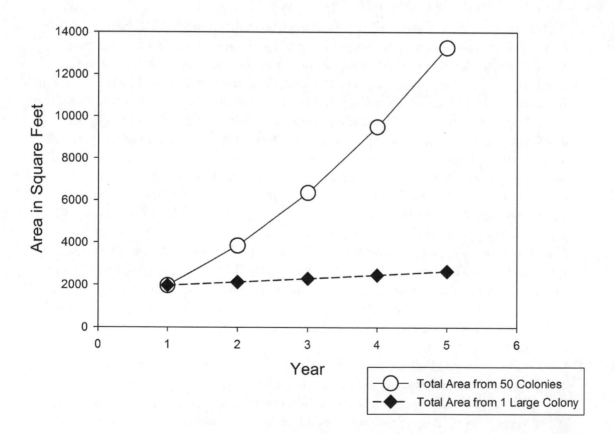

FIGURE 12-1. This graph illustrates the importance of controlling small colonies over a single, large core colony of exotic plants. Using an equivalent rate of annual spreading, the sum of the small colonies rapidly increases in the area occupied, far exceeding the area of the single, large core colony.

The area in the interior has little effect on the outward expansion of the colony but does contribute to the weed presence by seed dispersal. This is reason enough to attempt to reduce the main core population, in addition to the colonies, because this is the source of your satellite colonies.

If you have equipment at your disposal, such as bulldozers or tractors, then addressing these satellite colonies may not be an issue. Equipment can eliminate the small colonies very quickly, allowing concentration on the core area. Regardless of the equipment, however, it is important to focus on the satellite colonies first and then proceed to the main weed population.

Timing of Weed Control

Timing of weed removal can greatly enhance the overall impact your efforts will have on the weed population. As stated previously, whenever possible, invasive plants should be removed before they have a chance to set seed. Waiting until after seed has set will require aggressive soil treatment or manipulation.

A weedy tree species in the arid West of North America—the salt cedar originating from the Mediterranean—has seed that is viable for only one season. By eliminating new seedlings and eliminating the mature seed-producing plants, the salt cedar can be eradicated from a watershed essentially in one season, because there is no seed bank for the species. Numerous volunteers and volunteer groups have donated time and equipment to eradicate this plant from desert streams, knowing that timing is critical to the goal of salt cedar eradication. In many cases, following salt cedar removal, a dormant native seed bank has been able to germinate and reestablish a diverse riparian habitat. The key to these efforts has been understanding how the species exists in the landscape and using this knowledge as an advantage for the ultimate goal.

Weed Control Strategies at Ecological Restoration Sites

Ecological restoration areas generally require a more intensive weed control program than natural areas. Because of the predominance of weeds in most project areas, it is not expected that restoration areas will have 100 percent native groundcovers, even if native groundcovers are seeded. As described earlier, Priority 3 and 4 weeds may be left uncontrolled, at least in the short term, because they do not pose serious threats to restoration plantings or maintenance procedures, nor do they justify the cost of control efforts.

Weed control programs at ecological restoration project sites should call for complete elimination of Priority 1 species and control of Priority 2 species, generally requiring intensive aftercare for a three-year period. Control for Priority 2 weeds should be performed whether groundcovers were seeded one year in advance of tree and shrub installation, were seeded at the same time as tree and shrub installation, or were not installed at all. The approach, however, should be slightly different in each case, as explained below.

GROUNDCOVERS SEEDED ONE YEAR IN ADVANCE OF TREE OR SHRUB INSTALLATION

In this approach to restoration, the preferred groundcover species are seeded at least one year prior to the installation of trees and shrubs. During this first year, weeds that come up at the same time or after the seeded groundcover are controlled by mowing at the right time, with the right equipment, and in the correct manner.

Theoretically, this is the most logical method of weed exclusion, because the first year of weed control can be performed with a tractor-mounted mower, which can eliminate seed set of all Priority 2 weeds and many Priority 3 and 4 weeds as well. By mowing the site three to four times before weeds set their seeds, the native forbs and perennial grasses, previously installed by seed, are encouraged to spread. It is important to use a rotary or sickle bar mower and adjust the height above the grass culms (about five inches), so that the native grasses are not harmed during the mowings. Also during this first year, Priority 1 species that reproduce vegetatively should be hand-pulled or spot-sprayed as often as needed to eliminate them. This is important because the mowings will encourage these weeds to spread if they are not otherwise controlled.

During year two, when the trees and shrubs are installed, and during year three, specific control of individual weeds can easily produce nearly complete control of Priority 2 weeds. Several visits, however, may be needed each year due to the different times of germination, growth, flowering, and seed set. During years two and three, the site should be surveyed for Priority 1 species, and elimination performed if they exist. Thereafter, the trees, shrubs, and native groundcovers should be established enough to not require further control of Priority 2 species. Priority 1 species should still be surveyed for and eliminated if found.

Although this sequence of first planting and establishing a groundcover that can exclude weeds makes sense, it is not always feasible because of budget and timing constraints. The following situation is more typical.

Groundcovers Seeded the Same Year or One Year after Trees and Shrubs Are Installed

In this case, site preparation and preseeding weed control is even more critical than if groundcovers are installed one year prior to trees and shrubs. Because the trees, shrubs, and any containerized groundcovers, and possibly the irrigation system, will create obstructions, tractor mowing will likely not be possible unless wide spacing exists between plants. Some practitioners plant trees and shrubs in widely spaced rows to allow for the use of tractors and mowers. If the plantings will be irrigated, irrigation lines are installed along the rows of trees and shrubs to prevent interfering with the use of equipment. The area between these rows is generally seeded with groundcover after the woody plants and the irrigation system have been installed.

Specific control of both Priority 1 and 2 weeds should be implemented for three years, while the trees, shrubs, and groundcovers are becoming established. If adequate site preparation was performed and the seeding operation is successful, this task should not be excessive, although it will likely be more labor intensive than if the groundcovers were installed one year in advance. Both Priority 1 and 2 species should be individually controlled for three years, striving for complete elimination, especially for Priority 1 weeds. Thereafter, the restoration plantings and maintenance regime should not be hampered by Priority 2 species, and active control of these species can stop. Priority 1 species, however, should be surveyed for on an annual basis and eliminated if found.

Seeded Groundcovers Not Installed

If seeded groundcovers are not installed in the restoration area, it is generally for one of the following reasons. The first is in a case where the area is small and weed control can be performed by hand or an acceptable weedy groundcover exists. In this case, weed control can be limited to the three-foot zone around installed plants, with periodic mowing or cutting of weeds within the site if they grow tall and interfere with the success of the restoration or hamper maintenance procedures. Only Priority 1 weeds would need to be eliminated.

Another example in which groundcovers would not be installed in a restoration area would be where a thick organic mulch, such as straw, is used to suppress the weeds. This would be effective if the ground was weed-free before applying the mulch and the mulch was at least six to eight inches thick. In such cases, ideally one would plant native, rhizomatous groundcovers and encourage their spread as well as encourage the spread of any native groundcovers that "volunteer" on the site. The elimination of Priority 1 species would be necessary, although control of Priority 2 species would be optional.

Weed Control Methods

Weeds can be controlled by a variety of methods, including burning, flaming, mowing, hoeing, cutting, girdling, pulling, uprooting, tilling, smothering, mulching, applying herbicides, and installing weed mats. In the following discussion of weed control methods, a description of the three main groups of weeds (annual, biennial, and perennial) is also given because control methods differ for each group, except with the use of mulches and weed mats around individual plants (discussed last).

Annual Weeds (Annuals)

Annual weeds live no more than one year. They germinate, grow, flower, and set seed in a matter of months. Annuals generally reproduce by seed. Two broad categories are summer annuals, which grow in warm weather and survive as seeds during the winter, and winter annuals, which grow during the cool season and lie dormant as seeds during the summer. Annuals should be killed by cutting them off at least two inches below ground level by hoeing, pulling, and uprooting. They may also be killed by repeated mowings or cutting off aboveground. Mowing or cutting aboveground is most effective just as the plant is beginning to flower. However, if they recover, repeated mowing is required. Properly timed herbicide application can be highly successful without involving an extensive amount of repeated labor during the growing season.

Biennial Weeds (Biennials)

A true biennial plant germinates and grows in its first year and then flowers, sets seed, and dies in its second year. Biennials generally reproduce only by seed. Biennials sometimes act as annuals and flower in their first year (although they usually will survive their second year as well) or act like short-lived perennials, especially in warm climates. Biennial weeds may be controlled in the same fashion as annuals; however, they have greater recuperative power and are thus harder to kill or keep from reseeding. Therefore, for a sure kill, biennial weeds should be pulled or uprooted. Hoeing is not as effective because the root that is left in the ground may resprout, which would require repeat treatment. The more root system that is removed, the better is the chance of a complete kill. Mowing or cutting aboveground can be effective for controlling biennials, although it generally will not kill them, even if done during the early-flower stage. Repeated mowings or cuttings are necessary to eliminate seed set.

Primary Control Strategy for Annual and Biennial Weeds

The primary control strategy for both annual and biennial weeds is to not let them set seed. One plant can set thousands of seeds. The elimination of seed set in conjunction with an effective groundcover can control annual and biennial weeds in one to two years. The methods that kill annual and biennial weeds in one treatment (uprooting, pulling, and hoeing below ground) are preferred, especially if the targeted weeds do not form dense stands. This method ensures complete control without repeat treatments (at least on the same weed). Repeat treatments of the overall site may be necessary because of the staggered germination and growth of the weeds. Mowing or cutting aboveground should typically be used only if dense stands of targeted weeds exist.

As mentioned, several treatments may be required to keep the weeds from setting seed. Chemical herbicides are a cost-effective means of controlling annual and biennial weeds; however, they may disrupt soil microorganisms and mycorrhizal fungi, which are important for restoration plantings and the soil environment. Many weeds do not require fungus associations, whereas most native plants do.

Perennial plants live more than two years and may flower and set seed every year after reaching maturity, which may be the first season. Perennials commonly have extensive root systems. They may reproduce vegetatively or from seed. Perennial weeds generally can be effectively eliminated only by uprooting (best done when the soil is moist) or by spot applications of herbicides. Uprooting can be performed if the weed is relatively small or, if it is large, by using a brush extractor. Uprooting or pulling is the preferred treatment for plants that do not reproduce vegetatively.

Most perennial weeds reproduce vegetatively and thus are appropriately controlled using herbicides. Selected herbicides should be registered for use in or near waterways. Care should be taken to spray only the targeted weed. A fine spray is best, if it is not windy, to reduce the amount of spray that reaches the soil. Some perennial weeds are better killed using the cut-and-dab method of applying herbicide. In this treatment, the stem of the weed is cut and a strong solution of the herbicide is immediately painted on the cut surface.

To accomplish the maximum effect, apply herbicides at the appropriate growth cycle of the plant following herbicide instructions; this is usually when the plant is actively growing. The senior author advised two government agencies, one national and one local, of the need to apply herbicide at the correct time of the year (i.e., during its active growing period). In these cases, giant reed (*Arundo donax*) had already become dormant. The outcome was no kill, and the giant reed continued to spread the following growing season and outcompete the riparian species planted on the sites. The result was wasted money, wasted staff time, loss of habitat quality, and unnecessary introduction of an herbicide. Complete die-off is unusual, and in many cases repeat applications are required, especially for the hard-to-kill species. To succeed, a complete kill is needed on-site to be assured that the weed does not interfere with the maturation of the restoration site.

The primary control strategy for perennial weeds is to kill them outright. Secondarily, one should limit their spread by seed reproduction, if they reproduce by seed. It is always best to eliminate perennial weeds as soon as they are discovered because most of them rapidly spread into disturbed ground or even into fully vegetated land.

Mulches

Organic mulches can be used around individual plantings to suppress the germination of weeds. The mulch must be thick enough to exclude light from the soil, which sometimes necessitates reapplying additional mulch. Mulches do not work well on sites with steep slopes or on sites that flood periodically, or when using flood irrigation. Take care to ensure that the mulch is certified "weed-free." Noxious weeds have been spread to previously weed-free restoration sites through the application of straw mulch grown in fields invaded by noxious weeds. Weed-free straw is generally certified by the state department of agriculture in the state in which it is grown. Certified noxious weed requirements are set forth by the North American Weed Management Association. Rice straw is typically a good weed-free mulch and has been used frequently in riparian areas as well as in semiarid regions of the United States with great success.

Application of Herbicides

Pre-emergent herbicides can be applied to a site to prevent weed seeds from germinating. Post-emergent herbicides either kill the surface part of the plant (contact herbicide) or kill the plant by affecting the plant's biochemical pathways (systemic herbicide). Some herbicides are selective, meaning they kill only specific types of plants. Selection of the appropriate herbicide and the prescription for herbicide application should be made by an experienced herbicide applicator or pest control adviser.

The use of herbicides typically requires special permits depending on the ownership of your project site. Work on federal lands requires a Federal Pesticide Use Permit; work on state lands generally requires some sort of similar approval. All herbicide applications must be performed by a pest control adviser or someone with a qualified applicator's license. All herbicide applications should be documented and reported to the appropriate local and state agencies.

Weed Mats

Weed mats (sometimes referred to as fabric mulch) are pieces of landscape fabric placed around individual woody plants to exclude light and to prevent weed seeds from reaching the soil. They generally do a good job of weed suppression except when soil runs onto the mat, creating a germi-

> **Box 12-3. Lesson Learned: Temporary Nuisance Animal Control May Be Critical to Project Success**
>
> In the late fall, an organization installed numerous pole cuttings on a recently reconnected floodplain in an effort to create riparian habitat for native wildlife. That winter, the site was inundated with several feet of standing water for a couple of weeks. A site inspection that spring revealed that beaver had entered the site and girdled all of the pole cuttings at the water level, thus killing the trees. The beaver were native to the region and would be welcome inhabitants in the restored riparian forest. However, their early entry to the site resulted in replanting with added plant protection and animal control to allow sufficient time for the habitat to mature.
>
> The organization learned that during the plant establishment period and even into the stewardship phase, it may be necessary to exclude or manage nuisance animals even when they are native species.

nation site for weed seeds. Generally, a three- by three-foot square of landscape fabric is installed at each planting location and stapled to the ground on all four corners. Paper fabrics can also be used; however, they are generally good for only one year, whereas synthetic fabric lasts many years. Manual labor is involved in installing weed mats and removing them at the end of the plant establishment period. Hand weeding may still be required at the base of each plant.

Control of Exotic Pest Animal Species

In some cases, introduced or escaped animals pose significant threats either to native species or to the functioning of native plant communities. A common threat is the introduction of marine organisms from one region to another, either on the same continent or across oceans. Zebra mussels and freshwater clams are common in many parts of the world because of the extensive shipping now occurring (Cox 1999). During the early years of world exploration, goats, pigs, and other animals were introduced to many of the Pacific islands to provide food for subsequent travelers. These animals continue to seriously degrade many plant communities throughout the United States and have subsequently caused population declines of numerous species of birds, mammals, and reptiles.

Control of exotic animal species requires a pattern of approach similar to that for exotic plants; however, with the mobility of some of the species, it may not be feasible to have a methodical pattern of control. Several programs exist with the goal of trapping pest species, such as introduced red fox, Norway rat, and domestic cat. It may not be feasible to attempt to completely eliminate some exotic species; instead, the goal becomes to control the numbers to some reduced level so that native species in the degraded community can regain a presence at a level that permits them to reestablish ecological function (box 12-3). Consistency in a program is the key. This is especially true if it is beyond your ability to eliminate the source of invasion.

Chapter 13

Monitoring and Evaluation

Monitoring has one single important function: to provide information to enable an evaluation of the project as it is developing. Monitoring data should evoke some type of response, even if it is a "do nothing" response. This response will be an informed one and, ideally, one reached by considering all of the various factors affecting your project. Making intelligent decisions concerning any action on a restoration project requires a sound monitoring report.

Purpose of Monitoring

Although an important element for any restoration project, monitoring does not seem to receive the attention it deserves. All too often, projects have sufficient funding for the installation and the plant establishment period, but not for monitoring, because it is not considered as critical. We have observed several projects that have failed solely because of the lack of monitoring and a plan for corrective action when indicated. Nonprofit organizations and governmental stewardship agencies are constantly faced with the challenge of developing monitoring programs that will provide the needed data in a cost-effective manner. In many cases, this information can be useful for modifying future project design approaches or at least for making effective changes.

The primary purpose of monitoring a restoration project is to determine whether the project has met the objectives agreed upon by the project sponsors and stakeholders. The monitoring plan and performance measures are developed during the planning phase and refined through the design phase.

Regardless of the level of effort expended for your restoration project, a monitoring component should always be included in your management plan. We have identified three purposes for monitoring plans. First, monitoring of a restoration project is done primarily to document what is occurring on the project site. This is a way of assessing the progress of a project over time. Monitoring is helpful in evaluating the performance of specific techniques or materials (fig. 13-1) used on a site or a portion of a site (Clewell 1999). Various techniques and methods can be used on a project.

FIGURE 13-1. In this example of poor maintenance practice, a neglected tree has been permanently damaged by unnecessary staking and strapping that was tight and unattended. As a general rule, no plants should be staked; rather, nursery production should ensure that plants are free to react to winds and build reactionary wood. There are, however, exceptions when staking is necessary. In such cases, be sure to monitor plant condition to prevent damage. Double looping loosely with two stakes is best, if staking is required. (Photo by John Rieger.)

The effectiveness of a technique may require testing because not all techniques can be applied universally. Incorporating an experimental design into the overall project design can enable more rigorous evaluation (Kentula et al. 1993).

This leads us to the second purpose of monitoring: to establish a database for use in the planning of subsequent projects. Questions commonly arise related to the expectations of a planned project; having an existing database of prior projects and how they performed will provide confidence for the stakeholders and the team. Another benefit is to see how under differing conditions or circumstances, the response of the site will enable you to discard some techniques or adopt different material or methods that would normally not be considered for subsequent projects.

A third purpose is to verify that the project was executed as designed and is meeting the objectives or success criteria for it. These requirements may be ones that were generated for comparison purposes by you, the team, or the sponsor or applied by a governmental regulatory agency. Meeting or complying within these guidelines is often a requirement for permits.

Numerous methods and techniques are used in collecting data that describe the natural environment (Southwood and Henderson 2000; Manley et al. 2006). Knowing what data to collect is only the first step in a series of decisions you must make (Karr and Chu 1999). Depending on the questions being asked and the resources involved, specific methods are developed to best reflect the environmental condition. You should collect data that is directly related to the objectives and success criteria. The data should be collected at a level of precision that will provide reliable or consistent information that can detect any changes. These data sets will allow you to evaluate the status of the restoration.

Creating a Workable Monitoring Plan

A clear understanding of the project's goals and objectives is absolutely necessary in developing a monitoring program that will serve the stakeholders. Being able to demonstrate clearly that the objectives or trends are leading toward achievement of the project goals is an important element of the communication commitment to your stakeholders. It may be necessary to conduct some pilot studies to verify the ability to reliably collect consistent data, the proper use of any equipment, the time required to collect the data, and that the method is fully understood by all participants in the monitoring period of the project (Margoluis and Salafsky 1998).

The specific data to be collected and the frequency of collection have a direct bearing on the budget for monitoring. Table 13-1 lists attributes and characteristics that have been monitored in various projects, and also indicates recommended frequencies of sampling events. The monitoring frequency is ultimately determined by your goals and objectives, site conditions, and budget and by the desires or demands of your various stakeholders.

Ideally, development of the monitoring plans should occur after the objectives of the project have been finalized. However, in many cases, the desire to get on with the project takes over, and the work of sorting through all the monitoring issues becomes secondary, frequently personnel changes occur, or another organization assumes the monitoring following installation of a project. If you come on to a project without a plan in place, you will need to develop one immediately. The following list of questions was developed to assist not only those working on the project from the beginning but those coming into a project already well under way or in the postinstallation stage. The list is intended to help develop an efficient and effective monitoring plan that meets the established parameters for the project by the various sponsors and stakeholders.

- Why are you monitoring?
- What are the goals of the project?
- What are the specific objectives of the project?
- Is variation a desired objective?
- Have evaluation or performance criteria been established?
- Are the evaluation or performance criteria clearly related to the stated objectives?
- Are there criteria alternatives?
- Do the criteria incorporate natural variation?

Table 13-1. Compilation of Monitoring Characteristics, Attributes, and Frequency for Restoration Sites

Characteristic	Measuring	Frequency of Sampling
Hydrology		
	Salinity of water	Monthly
	Salinity of interstitial soil water	Seasonally
	Water levels at various tidal cycles	Spring tide cycle
	Tidal flow rates at distances from inlet	Spring tide cycle
	Tidal flushing maintained	Annually
	Streamflow rates	Seasonally
	Groundwater levels	Monthly or at least seasonally
	Flooding events	At occurrence
	Rainfall	Weekly during storm season
Topography		
	Elevations	Initially and after storm events or floods
	Slope of banks	Initially and after storm events or floods
	Configuration of special features	Initially and after storm events or floods after vegetation starts maturing annually
Soils		
	Texture	Initially
	Organic matter	Site evaluation stage
	Toxic substances	Site evaluation stage
	REDOX potential	Useful in diagnosing cause of plant mortality
Nutrient Dynamics		
	Inorganic N in sediments and pore water	Initially to plan for amendments; repeat if growth is poor
	Litter decomposition	Seasonal
Vascular Plants		
	Aerial photo of plant cover and habitats	Annually
	Heights and total stem lengths	Annually at end of growing season
	Vegetation growth/biomass	Twice yearly
	Cover of vascular plants	Annually
	Patch size or population of rare plants	Annually
	Density of annual plants	Annually
	Species diversity	Annually at various times of year to capture all species on-site
Consumers		
	Decomposers and shredders	Annually
	Aquatic insects	Seasonally
	Terrestrial insects	Seasonally
	Pollinators	Seasonally
	Predatory insects	Warm season
	Fishes	Annually or seasonally
	Birds	Weekly in nonbreeding season; correspond with migration if occurs in region
	Reptiles and amphibians	Weekly during breeding season
	Mammals	Weekly during breeding season

- Are the evaluation criteria driven by empirical data or regulatory requirements?
- Are the criteria developed from a reference site?
- What is the time frame for meeting the criteria?
- What is the basis for this time frame?
- What data are being collected?
- Are the collected data related to an established goal?
- What is the frequency of data collection?
- Is the frequency of data collection biologically fixed?
- What are the sampling methods?
- Have you familiarized yourself with the monitoring protocols currently being used by other restorationists at similar sites?
- Are the methods appropriate for the specific project?
- Are there more efficient means of collecting data?
- Can you use or slightly modify a standardized sampling protocol that will allow for comparison of your restoration project data with other restoration projects in the same type of ecosystem?
- Is it possible to reconsider methods, frequency, and quantity of data collection?
- How long is the monitoring period?
- Is the monitoring duration driven by performance or by elapsed time of monitoring?
- Is the project planned and designed to be self-sustaining?

Carefully evaluate the resource you choose to monitor. Generally, monitoring centers on vegetation development or biochemical condition of the soil and water, if present. However, if an animal species is selected that has not occurred on the site previously, you may be taking a gamble because you must rely on the suitable habitat being present at some time during the monitoring time period. A project that two of the authors were involved with required the verification of either the development of the vegetation community that met the reference database that was collected prior to project design or the documentation of nesting for an endangered species.

Several approaches have been developed to document monitoring, from simple visual inspection to detailed sampling over an extended period of time with subsequent statistical analyses (Krebs 1989). Once it is established what data are to be collected, there also needs to be consistency of how the data are collected. This is a critical attribute for monitoring; it will permit the comparison of data collected over time. The developer of the monitoring plan should keep in mind the ultimate goals of the project and use appropriate methods to ensure the conclusions are supported by the collected data (Margoluis and Salafsky 1998).

Details on the data sampling method, the data sheets to be used, and the statistical analyses to be used for the various data sets should be included in the monitoring plan. Numerous sampling methods and techniques have been developed to satisfy the needs of various research efforts. Site-specific requirements will determine which ones are suitable. Several techniques that reflect straightforward and efficient methods of data collection are discussed here.

Data Collection Methods

Numerous sampling techniques are available to use. Depending on the frequency and level of involvement required, you should try to select methods that are simple and direct and that can be done with a minimum of cost. Physical and biological resources will use different methods, adding to the complexity of the monitoring regime. Select carefully the parameters you need to monitor to meet your goals.

Physical Environment

Various wetland and riverine systems may benefit from collecting topographic data. The slope of channels, the length and shape of cross sections, and the presence and location of point bars and riffles are all features that affect the functioning of a watercourse. In an example from drier lands, Tongway and Ludwig (2011) conduct monitoring of elevations in Australian grasslands to control the distribution of water to maximize the goals of their restoration projects.

In wetlands, monitoring water chemistry is emphasized because it is a primary driver of the ecosystem. Data collection may include temperature, turbidity, pH, dissolved oxygen levels, electroconductivity, and nutrient levels (Howell, Harrington, and Glass 2012). The use of piezometer wells and gauges will provide the variability of water levels over time.

Biological Resources

Common plant and animal sampling methods include visual, transect, plot, point count, and territory mapping, to name a few. Detailed descriptions of these techniques have been published in various texts, field manuals, and ecology references (Bonham 1989; Elzinga, Salzer, and Willoughby 1998; Elzinga et al. 2001; Morrison 2009; Sutherland 1996; Southwood and Henderson 2000; Krebs 1989). Regardless of the method you choose to collect data, it is very important that the data collection be conducted as described for the method and that all the assumptions are met. In doing this, you will be able to apply statistical analysis to those techniques with quantifiable data.

Visual documentation is the simplest form of data collection. If the goal is simply to document presence or absence, then this method may be sufficient. A series of precisely located photo stations around the project site, with specific fields of view, can document not only the start of project installation prior to restoration but also the maturation of the vegetation over time. This method has limited potential for statistical analysis but has substantial value in preparing reports, presentations, and publications. Monitoring wildlife can be done with various photographic methods, including motion-activated cameras, infrared cameras, and timed-shutter cameras.

Sound recorders have been used to capture the sonar signature of bats as they fly overhead. These devices can also be sound activated, and with various electronic components, the data can be collected and downloaded after extended time periods.

Transects are a common method for data collection and can be applied to animals and vegetation communities. The data collected has the potential for use in a number of different statistical applications. A similar technique to transects is the line intercept method, in which plants are recorded where they cross a line. The total cover of the line is calculated in addition to determining species abundance and vertical stratification.

Plot method is a technique for sampling small plants, herbs, and grasses in discrete areas. Sizes range from 0.1 to 1 meter square or larger for woody plants. Plots are accurate for population counts.

Point counts and circular plots are bird census techniques commonly used on breeding bird populations, especially perching birds. Point counts require large areas so that point coverage does not overlap with adjacent point areas.

Territory mapping for birds and some reptiles and mammals can be encountered a number of times when conducting repeat site surveys. A composite map of these observations will generate a map of territories. Trapping, mark and recapture, catching, drift fences, and mist netting are several of the techniques used to verify the presence of small to large mammals, fish, and reptiles. Depending on the species, the trapping may be very time-consuming and may require significant equipment. State or province permits, certifications, or other authorizations may also be required.

Trails, tracks (e.g., footprints), antler marks, browsing lines, and scat (e.g., droppings, dung) are all good methods of determining presence. Under some conditions, and with appropriate planning on collection patterns, it is possible to apply statistical analysis to the data collected.

Monitoring Protocols

Standardized protocols have been developed for monitoring physical environmental conditions, vegetation, and wildlife populations. The use of standardized protocols is critical for comparing results across space and time. Protocols have been developed by various agencies charged with management of endangered species, species at risk, and rare, threatened, or sensitive species throughout North America. These protocols establish many of the elements previously discussed in a monitoring plan, augmented with specifics for each species of concern, such as time of day, length of sampling for that day, total sampling duration, frequency of sampling, and dates to sample within the appropriate season. In addition to these established protocols, the agencies require prior testing of the individual doing the sampling. You should familiarize yourself with the standard monitoring protocols being used in your region before designing your own monitoring program.

General Budgeting of Fiscal and Resource Needs

After you have completed the plant establishment period, you now have a persistent stewardship obligation. As discussed earlier, the obligation may be of varying intensity, seasonally or annually. Generally, the closer your project approaches self-sustainability, the less stewardship tasks you should have. Of course, if your target is a specific intermediate phase of an ecosystem or vegetation community, one that never reaches maturity, then your stewardship obligations will require action that keeps that cycle occurring constantly. An excellent example of this would be the heath of England and the prairies of mid-North America, where fire is frequently applied.

Projects done as mitigation frequently have a postinstallation requirement to manage the site for a number of years. We call this the habitat management period of stewardship, in which the site matures and develops additional habitat features through the natural processes of that vegetation community. One of the most significant activities during the habitat management period is monitoring. Projects done by nongovernmental and environmental organizations should make specific plans for conducting periodic monitoring. Monitoring is needed only until objectives or success criteria are met. Thereafter, "monitoring" becomes research for some purpose other than completing an ecological restoration project.

Because monitoring can pose a significant resource obligation over an extended time period, it is important during the planning and design stages of restoration to understand monitoring requirements and the methods to properly obligate resources and to identify equipment and supply needs. Having this information at the planning stage will properly inform the project stakeholders how the monitoring phase will be funded and executed.

The Elements of a Successful Monitoring Program

The level of experience of those taking the measurements, whether paid or volunteer, may convince you to alter the monitoring methods or sampling techniques. Usually, once the sampling regime and other questions have been resolved, monitoring consists of very simple measurements. These measurements can be done by volunteers, and they provide extremely valuable data that can be used in analyzing a project. Some techniques, however, may preclude the use of volunteers. Knowing these conditions will help you refine the monitoring plan so that you can make the best use of your resources.

Consistency Is Key

Consistency is paramount when conducting monitoring in which the data collected is intended to be statistically treated. Statistical tests have several assumptions, and these must be met to have valid conclusions. During the planning phase of the monitoring program, coordinate with experts to receive the guidance needed to ensure that the data are collected in a reliable and repeatable manner.

"Consistent" refers not only to the repeatability of the data collection method but also to when the data are collected. Sampling intervals need to be established, and the duration of sampling periods fixed. A significant factor in ensuring the consistency of data collection is to make the sampling techniques easily understood and transferable to others. Whether professional personnel or volunteers are involved, people come and go from a project. This reality needs to be considered from the start in developing the monitoring plans (fig. 13-2).

Relying on Volunteers to Assist in the Monitoring Activities

The limitations of volunteers fall into two primary categories: the first and most significant is consistency, and the second is performance. Will you have the same person doing the same task throughout the sampling period? If the sampling period is only for a few events, this may not be

FIGURE 13-2. Professionals and volunteers collect data on experimental plots studying the individual and combined effects of tree shelters, DriWater (gel-suspended water), and humic acid on plant survivorship and growth rates. San Diego, California. (Photo by John Rieger.)

a problem. However, a sampling regime may continue on a regular schedule for the life of the project. Dividing the sampling into elements or efforts that are not too demanding can yield better results. Varying tasks is also needed because people have other commitments and cannot always comply with the schedule you have developed.

It is important to have clear and concise instructions on sampling techniques. Giving workshops or other types of training programs will minimize the variability due to different tasks. This situation is not unique to volunteers. For this reason, it is a good practice to develop a techniques manual with full explanations of the sampling methods and where and how the various sampling locations are or will be determined. It is difficult to fully explain the monitoring operation in a contract, so be sure to include conditions that permit review and modification or a trial period to ensure that the monitoring is progressing as you envisioned.

Maintaining Good Monitoring Records

Regardless of who conducts the monitoring, all records should be kept in a consistent format, in one location, and kept current. All data forms or pages must be dated (day-month-year) and initialed by the recorder. This will allow the information to be accessible to others as well as to any regulatory or oversight organization. Often neglected but nonetheless important is the publication of results. Published results advance the practice of restoration by sharing successes and failures and are even more valuable when the data collected meet the rigors of statistical analysis, thus allowing comparison by others with similar or contrasting conditions or sites.

Research Advisory Committee

Large, complex, and environmentally or politically sensitive projects may benefit from creating a research advisory committee. This committee can assist in the monitoring aspects as well as facilitate communication between research institutes and land managers. The committee also enables practitioners to explain management priorities and the acceptable adaptive management strategies, allows for peer review and quality control, reduces redundancy, assists in developing research priorities, and increases the links between research and applied science.

Evaluation

The whole purpose of monitoring is to collect data that can be compared or judged against the reference or some established criteria. Monitoring can be conducted prior to initiating the project to document the existing site conditions. We have monitored groundwater levels over a year's span to understand how the water levels responded seasonally and to provide us with a reference for determining elevations of the restored site. Monitoring of water chemistry is one of the more common data sets collected for restoration projects, in addition to presence/absence data for various forms of wildlife. Preproject monitoring can be a valuable task because it provides needed information to compare with the postinstallation project. Poplar Island, Maryland, in the Chesapeake Bay, is an example of a complex project with multiple factors that may affect the outcome of the project. Preconstruction monitoring was conducted to have a reference from which to compare effects of the project following construction and plantings (box 13-1). Preproject monitoring can also be used to convince people that the site really was degraded prior to your successful project.

If properly formulated, the data collection regime will address the specific characteristics of the project site that allow an independent comparison or assessment of compliance or attainment to a specific criterion or set of criteria. Variances in the soil chemistry, genetic material differences, or predation on seeds or plant material may alter the growth performance. These and a wide range of other manifestations can, and often do, occur. Sometimes, this lack of performance may not be obvious and is only detected after analysis of collected data. After it is discovered, it becomes important to involve the appropriate team members to discuss the possible ramifications.

Tongway and Ludwig (2011) provide several excellent examples on how the results of monitoring led to responses for corrections or reconsideration by their stakeholder sponsors. It is only by monitoring and then evaluating the data that remedial actions can take place that will yield the desired results. Another situation may be that an unanticipated event occurred and the outcome is a normal result. In many cases, these single events will correct themselves with time, and a "do nothing" approach may be appropriate. Alternatively, sometimes a reason for the lack of performance is not known. It is entirely possible that it will require detailed investigation (e.g., soil chemistry analysis for heavy metals) that was not executed during the preproject inventory. Whatever the plan of response, it is important that it be documented fully with the reasoning, or the response action and any requirements, along with a schedule and further discussion as necessary. In the final analysis, natural systems, whether human placed or spontaneously occurring, will behave differently because each site is unique. Examining the data collected to establish trajectories will provide additional information that may guide your response in a different direction. Dare we say that you should exercise "common sense"?

Box 13-1. Restoration Project Highlight: Monitoring Data Allows Project Managers to Modify Future Phases of a Large Project (Adaptive Management) to Improve Overall Project Success

Location: Paul S. Sarbanes Ecosystem Restoration Project at Poplar Island, Maryland, United States

The Paul S. Sarbanes Ecosystem Restoration Project at Poplar Island can be considered a model for beneficial use of dredged material. An extremely large undertaking, the project involves the cooperation and partnership of eleven different organizations, including federal agencies, state departments, and two universities. The principal funding organizations and in the role of project management are the US Army Corps of Engineers, Maryland Port Authority, and Maryland Environmental Service.

Located in the Chesapeake Bay seventeen miles south of Annapolis, Maryland, the project seeks to "restore island habitat to the Chesapeake Bay" with the goals of creating remote and diverse island habitat, restoring quiescent water habitat in Poplar Harbor to promote submerged aquatic vegetation recovery, creating/enhancing tidal wetlands to provide fish and wildlife habitat, and creating a bare or sparsely vegetated island within the restored marshes to provide nesting habitat for birds

The ultimate plan for Poplar Island is to create 1,140 acres, half wetlands and half uplands (570 acres of each), which approximates the configuration of the island in 1847. The wetlands will be primarily in the eastern half of the island, and the forested and meadow uplands in the remaining area. The marsh will be a low marsh at 80 percent of the 570 acres, and the remaining 20 percent will be high marsh. The island will have a capacity to receive 40 million cubic yards of clean dredge material from the Baltimore approach channels.

The most recent cost estimate to completely create the island is 1.2 billion dollars, with a completion year estimate of 2041. Initial construction began in 1998, with exterior dikes of the island reaching ten feet above mean low water by 2001. The island has been partitioned into approximately forty-acre acre cells for dredging and subsequent habitat restoration activities. This method will permit a mosaic vegetation pattern and also create habitat in parts of the island prior to receiving all of the dredge material.

Water quality monitoring was conducted during the preconstruction activities (1995–1996), and turbidity monitoring was conducted during the phase 1 and phase 2 perimeter dike construction at Poplar Island (1998–2001). Turbidity standards were in effect during construction, and continual monitoring assured compliance. Typical parameters monitored include turbidity, salinity, conductivity, temperature, dissolved oxygen, total suspended solids, and nutrients (nitrogen and phosphorous). In addition, the monitoring includes analysis of chlorinated pesticides, polycyclic aromatic hydrocarbons (PAHs), and polychlorinated biphenyls (PCBs).

As new habitats are created, the plants and animals within the habitats are monitored to evaluate habitat function. Habitats are directly related to their location in reference to the water level. Within the low marsh, islands are formed to add diversity in the edge configuration of the marsh and the berm that surrounds Poplar Island. Outside of the berm is a rocky shoreline and shallow water. The tidal changes provide a variety of water depths, mudflats, and shorelines for numerous forms of animals. Particularly important is the habitat that enables submerged aquatic vegetation to persist. This is considered a critical Chesapeake Bay habitat. Submerged aquatic vegetation has a significant ecological role in the bay by providing food and habitat for waterfowl, fish, shellfish, and invertebrates. Also, submerged aquatic vegetation provides nursery habitat, produces oxygen in the water column, filters and traps sediment, and protects shorelines from erosion by slowing down wave action.

Mammals, reptiles, and amphibians were surveyed in 2002. The mammals included beaver, white-tailed deer, river otter, raccoon, and house mouse; these species were not introduced by the project but came from nearby islands. The brown water snake, diamondback terrapin, and American toad were also recorded. The presence of terrapin is especially important because their habitat on the mainland shoreline has been significantly impacted by development. One hundred terrapin nests were found on Poplar Island. This surprising and very welcome discovery required the addition of an aluminum fence to prevent terrapins from entering cells still under construction.

Control of invasive and nonnative species to the island will be needed for the project to succeed. Most notably already requiring control are the common reed (*Phragmites australis*) and the mute swan (*Cygnus olor*).

Wildlife management activities that have been initiated include habitat enhancements, such as shrub plantings and the placement of Christmas trees and snags. An overall wildlife management plan has been developed to guide the activities into the future. Nesting structures have been installed to encourage nesting of osprey and tree swallows as well as other species requiring larger platforms or cavities. The control and surveillance of predator and nuisance species, as well as a program of monitoring of wildlife and wetland development, have been initiated.

PART V

Synthesis of the Process

We hope that this book will take most, if not all, of the mystery out of how to do an ecological restoration project. Although not all projects are alike, the process we have discussed throughout can work with all projects. We have included a small project that we feel demonstrates the use of our framework in action. We end with a review of what we consider the important project planning and management principles pertaining to ecological restoration. We cannot overemphasize the importance of becoming part of the ecological restoration community. Exchanging experiences and learning from and informing others are the best ways to learn and develop the practice.

Chapter 14

Bring It All Together

In this chapter we bring all of the previous discussions together, walking you through the major steps in the restoration planning and design process (fig. 14-1). We have included several sample documents from one of our completed restoration projects. These are excerpts from the project file, with notes attached for clarification. Where the plan sheets were too large, we have included only a portion of the plan sheet to demonstrate a technique or to point out a specific feature.

Project Background: Los Peñasquitos Canyon Preserve

In the early 1990s, two of the authors, then working for the California Department of Transportation (Caltrans), developed a biological mitigation plan for a nearby freeway project. The mitigation requirements called for the creation of three acres of southern willow scrub vegetation. As the project sponsor, Caltrans was obligated to perform the mitigation. The authors developed a restoration project to satisfy the mitigation plan set forth by the regulatory agencies. A

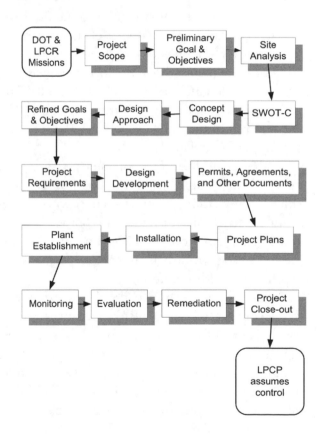

FIGURE 14-1. Flowchart of the Los Peñasquitos Canyon Preserve restoration project.

217

site suitability analysis was conducted, and several candidate sites were investigated. Regulatory agency staff helped the project team identify a suitable area within the Los Peñasquitos Canyon Preserve (LPCP) where the mitigation project could be undertaken.

Working with the park owners, the County of San Diego Department of Parks and Recreation, along with the City of San Diego, regulatory agencies, and local community planning groups, the authors developed a comprehensive plan for the restoration of a three-acre plot in the heart of the preserve. Los Peñasquitos Creek runs east to west from the interior of central San Diego through LPCP, an open space park of approximately 3,400 acres located in the northern part of the city of San Diego. The County of San Diego is the primary owner of the preserve and is responsible for its operations and maintenance.

LPCP was once a thriving cattle ranch, which first came into being in the 1860s. An original California Spanish land grant, the ranch soon grew from its original 4,230 acres to more than 14,000 acres. Primary agricultural activities included livestock (primarily cattle) and some grain crops. Historically, Los Peñasquitos Creek supported a diverse and thriving riparian habitat. Dominated by southern willow scrub vegetation, the creek bed meandered along the canyon bottom as it ran beneath shady canopies of large sycamores (*Platanus racemosa*), coast live oaks (*Quercus agrifolia*), and cottonwoods (*Populus fremontii*). The canyon slopes were covered with thick chaparral scrub on the north-facing slopes and with coastal sage scrub on the south-facing slopes. By the time the cattle ranch operations got into full swing, much of the native canyon bottom vegetation had been altered, through clearing to allow cattle more grazing and access to water, and through the removal of the native vegetation along the banks of the creek and adjacent canyon bottom, apparently for firewood.

In the 1960s, the County of San Diego acquired Los Peñasquitos Ranch (fig. 14-2); in the 1970s, the Los Peñasquitos Canyon Preserve was established. The mission of the preserve is to protect and enhance the natural resources in the canyon and to protect the historic core of the ranch as a museum of early California ranch life. Some goals of the preserve are to restore the natural vegetation and to maintain recreational opportunities for hiking, cycling, jogging, and equestrian uses. The preserve also provides an educational forum for elementary schoolchildren to learn more about ranch life in early California.

In 1987, county planners developed a draft master plan for the preserve to provide the overall framework for the County Parks Department to protect, manage, restore, and enhance the biodiversity of one of southern California's last undammed canyons. In addition to the goal of enhancing the natural resources, including native vegetation, wildlife, streams, and canyon walls, the draft master plan specifically targeted the period of 1862–1872, the time the ranching operation was just beginning. This goal applies to work within the historic core of the original ranch buildings (all adobes) and the canyon vegetation itself.

Project Scope

Thus, our project scope was to restore three acres of southern willow scrub on Los Peñasquitos Creek and establish riparian vegetation along the canyon bottom as it appeared during the peri-

Figure 14-2. The restored historic Johnson-Taylor Adobe ranch house, built about 1860. Los Peñasquitos Canyon Preserve, California. (Photo by Mary F. Platter-Rieger.)

od of 1862–1872. The "'historical reconstruction'" design approach required the removal of three large groves of nonnative eucalypts (*Eucalyptus* spp.), which had come to dominate the canyon floor within the project area. The project team used archival photographs and lithographs, as well as historical information provided by preserve staff and the Friends of Los Peñasquitos Canyon Preserve (FLPCP). The historic photographs illustrate that the site was occupied by riparian vegetation, including sycamores, cottonwoods, mulefat, and oaks along the banks of the creek. In addition, the project team made use of field survey records from other nearby streams.

In return for using parklands to mitigate a public project, the County Parks Department also required that the project team remove several volunteer palm trees that had grown adjacent to and in the historic Spring House (fig. 14-3), which were beginning to undermine the foundation of the house. However, five historic Phoenix palms planted at a nonthreatening distance from the Spring House would be preserved.

FIGURE 14-3. Now restored, more than forty volunteer palm trees were damaging the historic Spring House. Removal of these palm trees was a project requirement placed by the County Parks Department. San Diego, California. (Photo by Mary F. Platter-Rieger.)

The landscape architect prepared the project plans and bid package in consultation with the Caltrans biologist. The Caltrans construction division provided the contract administration for the project installation, and the project was installed by a contractor. The documents prepared and major events for this project are listed here:

1. Site analysis, March 1991
2. Conceptual design, April 1991
3. Project plans, June 1991
4. Begin construction, September 1991
5. Plantings and seeding installed, January 1992
6. Installation of remaining features complete, May 1992
7. Streambed agreement for California Fish and Game, compilation November 1994
8. Annual monitoring reports, October 1992–1994
9. Complete plant establishment, May 1995
10. Remove all aboveground project features, July 1995
11. Project closeout, November 1995

Goals and Objectives

The goals of the project satisfied the mitigation requirements for a transportation project. The project team identified three specific restoration project goals as well as several objectives for each:

Goal 1: Create three acres of self-sustaining southern willow scrub habitat similar to that which existed at Los Peñasquitos Rancho in the early 1860s.

Objectives:

1-01. Remove all nonnative species from the three-acre site prior to planting trees and sowing seed.
1-02. Completely remove the top 6 to 8 inches of soil and leaf litter from the site.
1-03. Start tree planting on three acres by October 15.
1-04. Plant three acres of southern willow scrub in Los Peñasquitos Canyon Preserve as identified on maps.
1-05. Use fencing and/or plant protectors on all tree plantings as needed to ensure protection from herbivory and vandalism.
1-06. Remove all physical evidence (fence, irrigation, and so forth) from the site within three years of project installation completion.

Goal 2: Protect historic features within the project area.

Objectives:

2-01. Avoid any contact with the five historic Phoenix palms near Spring House.
2-02. Maintain functional use of park trails and minimize interference of users in and immediately adjacent to the three-acre site during project installation and monitoring.
2-03. Establish a path for heavy equipment that prevents vibration-based impacts to the adobe buildings.

Goal 3: Protect existing native vegetation within the project area.

Objectives:

3-01. Prior to the beginning of work, identify construction staging and storage areas on-site and clearly delineate their boundaries.
3-02. Prior to initiating a contract, produce a detailed native vegetation map of sites and locations and delineate with flagging or protective fencing within the construction zone and along all access roads the contractor will use.

SITE ANALYSIS

Members of the project team conducted a detailed site analysis. Prior to the actual fieldwork, the team members studied aerial photos, topographic maps, utility plans, ownership records, and previous reports obtained from the County. Combined, these documents provided valuable insights into the history and various issues surrounding the project site. Observations from the site analysis were recorded using the site analysis checklist presented in appendix 10. Using the information in the checklist, a concept plan was developed (fig. 14-4). The site analysis checklist provided an opportunity to record both observations and any action items resulting from the issues presented in the field.

SWOT-C ANALYSIS

During the site analysis, preliminary assessments were made and the site analysis checklist items were categorized according to their relative strengths, weaknesses, opportunities, threats, and constraints. After the site analysis fieldwork was complete, the project team summarized the findings and developed a list of the SWOT-C items:

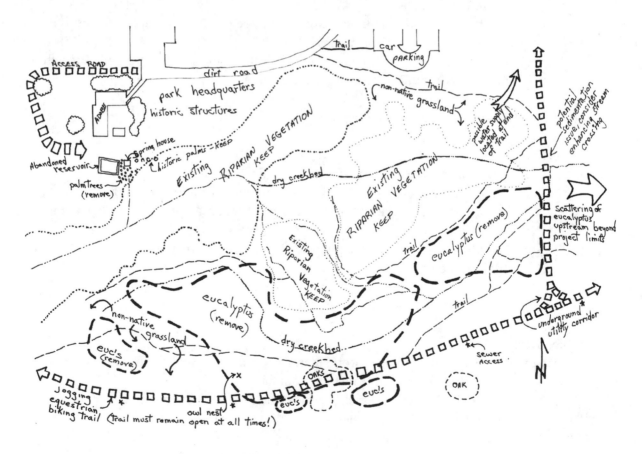

FIGURE 14-4. Site analysis map of Los Peñasquitos Canyon Preserve.

Strengths
- Remnant population of Platanus, Quercus, and Baccharis
- Existing creek flows year-round
- Wetland soils intact
- Canyon walls buffer creek area from adjacent residential areas
- Active owner/manager/steward of site by the County Parks Department

Weaknesses
- Nonnative monoculture (eucalyptus trees) dominating the creek bottom up and downstream: several scattered throughout the project area
- Absence of native understory along creek banks
- High pedestrian, mountain biking, equestrian trail users
- Large volume of eucalyptus leaf litter and seed bank covers soil surface
- Volunteer Phoenix palms near Spring House crowding out native vegetation

Opportunities
- Existing base of committed volunteers—Friends of Los Peñasquitos Canyon Preserve
- Committed stakeholder
- Control unauthorized trail development
- Return cottonwood species to drainage that was removed during ranch operation
- Draft master plan for Los Peñasquitos Canyon supports proposal
- Existing intermittent creek channels parallel to Los Peñasquitos Creek

Threats
- Series of trails for equestrians, hikers, mountain bikers
- Flood potential prior to plants being fully established
- Five historic Palms to remain, represents seed bank and potential for future problems
- Vandalism potential is high due to persistent park use
- Surrounded by urban development
- Utility corridors and serving utilities prior rights, potential for disturbance to project area with major utility maintenance activities
- Potential erosion and sedimentation on Ranch House trail over creek

Constraints
- Protect all historic features within project area
- Protect native vegetation within project area
- Owl's nest active
- High volume of daily park users
- Utility corridor present with requirements for uninterrupted access

PROJECT REQUIREMENTS

The project team synthesized the SWOT-C analysis and the project objectives and translated these items into specific project requirements. Additional project requirements resulted from discussions with County Park staff. Nonnative grasses between the creek and the ranch house road were to be removed and replaced with native wildflowers. Access into and through the project site was restricted to prevent damage to the historic structures.

Other project requirements included the following:

- Plant establishment for a three-year period to ensure self-sustaining vegetation
- The relocation of an owl nest from an eucalyptus tree within the construction zone into a nearby sycamore tree outside the construction zone
- Removal of project-installed fencing following successful plant establishment
- Maintenance of service road access at all times during construction

The resulting requirements then became the key drivers behind the design development process.

Design Approach

The County's draft master plan identified the need to restore the riparian area to the condition as it appeared between 1862 and 1872, when the ranch was just beginning operations. Therefore, the historical reconstruction approach was determined to be the appropriate design approach for this project. The development of the concept design relied on references from many sources, including historic photographs of the ranch, drawings, and remnant patches of native vegetation (in and near the creek), and journal notes recorded by the original ranch owner. Members of the FLPCP, a nonprofit volunteer organization that supports the preserve, provided information and insight into the canyon's ecosystems, which was extremely valuable. The FLPCP provided detailed information concerning the hydrologic cycle as well as specific information about wildlife movements that the authors would not have gathered without much longer periods of site analysis and monitoring. The combination of these various data sets was used to develop the reference model that generated the design.

Concept Design

Major project features were identified on the concept plan, which was prepared based on the data gathered from the site analysis. This concept drawing (fig. 14-5) became the basis from which the project plans were prepared during the design development phase.

Design Development

Preliminary planting plans were generated from the concept design (fig. 14-6). Species composition, desired spacing, and plant container sizes and types were determined based on the restoration planting model generated by the biologist. This model was originally developed for use on a similar restoration project. The basis of the model was an extensive field analysis of vegetation

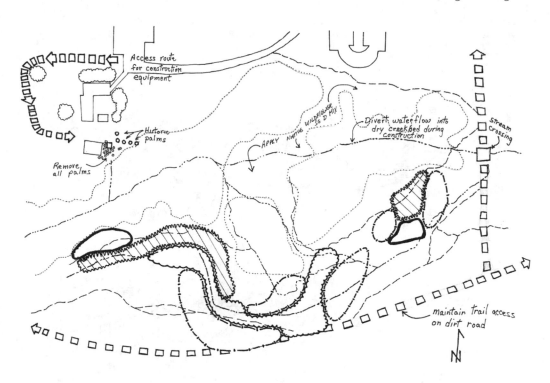

FIGURE 14-5. Bubble diagram of Los Peñasquitos Canyon Preserve design concept.

of similar habitat type occupied by an endangered bird species. The vegetation model, budgetary constraints, and anticipated environmental stressors during plant establishment influenced container size and quantities. Therefore, densities and sizes were the result of the model and of site-specific mortality expectations.

Permits, Agreements, and Other Documents

Permits and agreements covering streambed alteration (US Army Corps of Engineers 401/404 and California Department of Fish and Game Streambed Alteration Agreement 1601) were obtained from the state and federal agencies having jurisdiction. In addition, several local approvals were needed to work within the county park. Approvals were obtained from six local community planning groups and from the Los Peñasquitos Canyon Task Force, composed of various governmental representatives. The FLPCP again provided valuable support for the project by speaking in favor of it before several committees, planning groups, and other governmental bodies, and by writing news articles.

Project Plans

Contract drawings and plans (chapter 9) were prepared based on the preliminary planting plan. These contract drawings provided accurate (to scale) depictions of planting areas and exact quanti-

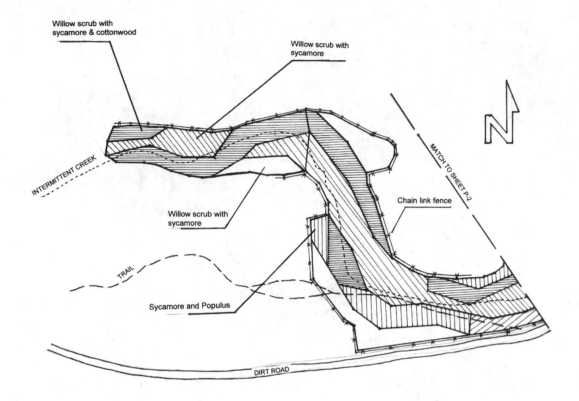

FIGURE 14-6. Preliminary planting plan for Los Peñasquitos Canyon Preserve.

ties of plant species and sizes for installation. Additional project features—such as barrier fencing, access gates, access control, and utility points of connection—were all shown or noted on the drawings. Other contract drawings—irrigation, clearing (demolition) plans, construction details, project location map, and plant list specifications—were also combined to create the design drawings. These drawing plans became the basis for quantity calculations for all construction material: plants, irrigation equipment, fencing, interpretive signs, and seeds.

Quantity calculations were totaled, and unit costs for each item were determined based on prevailing cost data in the county. The resulting bill of materials became the basis for bidding on the contract and was used extensively during contract administration to facilitate the negotiations for contract changes. The project designers prepared specifications with the assistance of the project biologist and project construction administrator. These specifications described the requirements regarding materials and workmanship. They did not describe how the work was to be completed; rather, they described the expected product or deliverable.

These three documents—project plans, bill of materials, and specifications—were then assembled into a bidder's package and circulated to contractors. The bidding period was approximately four weeks. Once the low bidder was identified and an agreement reached, the contractor began work in September 1991.

Installation

The Caltrans contract administrator, in close coordination with the project biologist and the landscape architect, administered the contract to ensure the contractor delivered the project as planned (fig. 14-7). The contract administrator relied on daily field inspection to ensure compliance with the drawings, BOM, and specifications. Where deviations were noted, the contractor was issued notices to comply with the project plans. If compliance was not achieved, monthly progress payments were withheld. The project biologist and landscape architect were consulted regularly for additional support to process changes that occurred because of unforeseen circumstances in the field. This project had several infrastructure elements within the project limits. Some of the information on location, size, and quantity received was inaccurate, requiring adjustments.

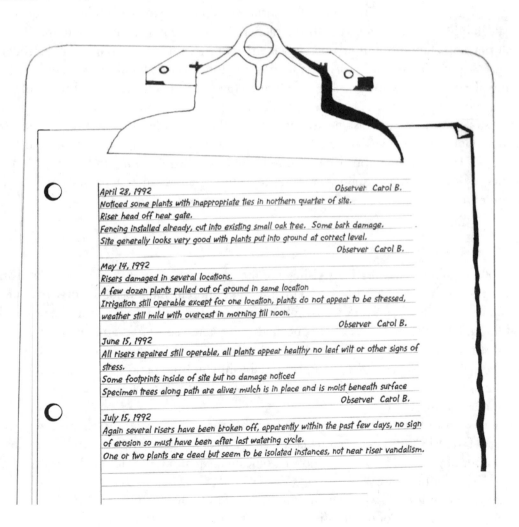

FIGURE 14-7. Monitoring field notes for Los Peñasquitos Canyon Preserve.

The project began prior to the rainy season with the eucalyptus removal and the clearing of leaf litter and seeds, which was completed within about six weeks. These activities gave the appearance of a major logging operation. The wood was cut to fireplace lengths and made available to the local community. This helped to reduce costs of hauling and disposal fees and also created additional goodwill with nearby residents. Leaf litter, seeds, weeds, and other trash and debris were removed and hauled to a local landfill.

After the clearing (demolition) and soil removal process was completed, irrigation work was allowed to begin. PVC pipe was placed twelve inches below the surface. Rotary sprinklers mounted on high, two-foot risers were used to provide supplemental water during the first two years of plant establishment. A solar power panel was installed to provide electricity to operate the controllers and timers for the watering cycle.

When irrigation work was completed in one area, planting immediately began in order to complete all planting prior to the winter rains. Prior arrangements were made with local vendors to ensure the availability of plant material to be planted at the sizes required. Applicable specifications previously developed for restoration projects were used to ensure proper treatment and installation of the site. We provided on-site inspectors to examine the plant material for general health and vigor of the plants as well as any possible infections. This construction staging technique of having concurrent activities is known as fast-tracking. The project installation was completed prior to the first winter storm. Without the fast-tracking technique, this schedule would not have been met. A six-foot-high chain link fence was erected around each planting area, with a three-foot-wide access gate. Placement of signs at various locations, removal of an illegal mountain bike creek crossing (a preserve-requested action), debris removal, and other maintenance activities completed the installation effort.

Monitoring

Members of the project team regularly visited the site after installation (fig. 14-8), and observations on the progress of the site were made and documented in field monitoring reports. The project was considered to be small, and it was possible to monitor all the plantings within the project limits. In addition, all of the locations where eucalpyts were removed were examined for resprouting. Data were simply tabulated as individual mortality totals. The number of dead plants was very low, and these were replaced during the plant establishment time period. Remediation issues involving irrigation system repairs and replanting of dead plants were noted and discussed with the construction administrator.

Several acts of vandalism were recorded during the first year; many of these reports were made by members of the FLPCP. These acts usually resulted in damage to sprinklers and the removal of a small number of plants. The timely reporting of vandalism allowed us to repair damages before secondary effects could occur.

The contractor performed weeding on bimonthly intervals during the first year. This aggressive effort was required because of the invasive nature of the eucalpts and palms previously occurring on the site, as well as other weed species opportunistically exploiting the newly available habitat.

FIGURE 14-8. Cleared of eucalypts, this site was planted with container-grown willows, cottonwoods, and sycamores. Water was returned to the original streambed following completion of planting. Photograph taken four months after installation. San Diego, California. (Photo by John Rieger.)

Control measures included hand pulling and a limited application of herbicides in more concentrated population areas.

Annual evaluations were conducted based on the field monitoring reports. These evaluations were done to assess trends in site conditions and were used as a basis for remedial action. Replacement plants were installed in areas where mortality exceeded expectations. Experience from several prior projects, in similar habitat and region, formed the basis of our expectations. The project plant survival numbers were compared with the success criteria from previous projects. Although mortality and vandalism created some challenges, the project was very successful overall. The plantings performed extremely well with mortality not exceeding 5 percent, well within our anticipated mortality range of 20 percent.

Project Closeout

As-built drawings were prepared to document changes to the project design made during construction and plant establishment. These as-built drawings were transmitted back to the project team for reference and storage. All additional project documents, including permit compliance, reports, monitoring reports, and evaluations, were assembled into a project history file and transmitted to Caltrans's Record Center for storage. A final expenditure report was prepared to determine total project costs. These costs were compared with the budget, and a report was submitted to the project sponsor. A final report prepared by the contract administrator noted deficiencies in the project plans, highlighted contract change orders, and included recommendations for future projects. This report was submitted to the project manager and distributed to the project team. The project concluded with the submittal of this report.

Project Epilogue

The site, plantings, leaf litter accumulation, and wildlife use continue to develop as anticipated. The success of this first restoration project on Los Peñasquitos Creek prompted the volunteer group, the FLPCP, to develop a weed removal program for areas surrounding the project site. The FLPCP targeted remaining eucalyptus trees within the preserve—an approximate nine-mile reach of Los Peñasquitos Creek. Although the initial restoration project eliminated the most concentrated population, there were several scattered colonies throughout the preserve. Many of these groupings were much smaller and were within the limits of handwork done by volunteers. The FLPCP also has been active in removing other weed populations, greatly enhancing the biological values found within the preserve.

Several interesting events have been noted during subsequent visits to the site to observe the process of maturation. Mortality of plants of all species continues, presumably as a result of competition and one not unexpected as a natural process in the maturation of riparian vegetation. Coverage continues to be within the expected range. As plants die, others adjacent to the clearing increase their spread. Leaf litter has slowly built up, and various herbaceous plants have sorted themselves into patterns. It is not clear what resources are prompting the patterns. Minor erosion patterns were noticed, largely the result of concentrated water flow. Flooding of the site has taken place but not at the frequency anticipated; general rainfall in the region has been atypical to previous seasons. After twenty years, no signs of overuse are present in the understory (fig. 14-9). The site remains on a successful path of maturation.

The LPCP project illustrates how this framework can be of assistance in the planning and execution of a restoration project. In many respects, the LPCP project was simple and small in size. However, it was not without its challenges. We had a strong project management team of co–project managers. Using our framework, we were able to quickly develop a project with very little rework and a minimum of delay. We had not expected to conduct the numerous public meetings, and we received some opposition. Our close relationship with the major stakeholders was instrumental in moving the project forward. Had we not nurtured this relationship early in

FIGURE 14-9. Another portion of the restored site twenty years after installation; cottonwood and sycamore exceed thirty-five feet. Several flood events have sculpted the riverbed, and a complex understory habitat has developed. San Diego, California. (Photo by Mary F. Platter-Rieger.)

the design process, we could have had some delays conducting orientation meetings to bring all of our stakeholders up to date.

Following the methodology for defining a restoration project presented in chapter 3 was also instrumental in allowing us to move forward with a minimum of delay. We did discover some issues that were not readily apparent upon initial site inspection. By following the site analysis checklist, we were able to not only identify these issues but also develop opportunities resulting from a SWOT-C analysis. During aftercare, the site experienced a flood that caused some damage. Already in place were requirements for replacements and repair for a one-year period. This we felt was enough time to have the plantings become fully established to withstand flood flows, which are periodic and variable in this drainage.

We strongly believe that the framework presented in this book will greatly benefit any restoration project when applied appropriately. In other words, not everything in this book will apply to every project, so it will be up to you to decide which specific items need not be of concern. We do suggest that you at least review each item before you dismiss it as not applicable. By following those aspects, backtracking or "redos" should be significantly reduced if not eliminated. Our goal in writing this book is to improve the likelihood of successful projects. Following our framework through the project phases will lead to that success.

Chapter 15

Synopsis

By reading this guidebook, we hope that you have learned a great deal about how to plan and carry out an ecological restoration project. The framework and tools we have provided should aid you in achieving a successful restoration project.

Project Planning and Management Principles

The following summarized principles will assist in your recall of the planning steps. This list can also be of help when educating your stakeholders, team members, supervisors, or funding organizations. Adhering to these restoration project planning and management principles is the key to achieving a successful restoration project, on time and within budget.

1. *Prepare a plan and go through a planning and design process tailored to the needs of your project.*

The overall process of developing a restoration project involves proceeding through four phases, the first two of which are planning and design. Within each phase, you will need to proceed through a series of steps before you arrive at a well-conceived project to achieve your stakeholders' vision of a restored site. Simultaneously, you will be evaluating the application of one or both strategies, management and construction/installation, for implementing your project.

2. *Engage all stakeholders and interested parties in your planning process from the beginning.*

The involvement of all stakeholders in the project planning and design phases is critical to the long-term success of your project. The term *stakeholders* includes not only the project sponsors but also those who will benefit from your project as well as those who might otherwise be affected

233

by your project. The involvement of regulatory agency staff who will issue permits for your project work in the planning process can smooth the way for a more efficient permitting process.

3. *Fully investigate the current and historic conditions on your project site and in the surrounding area, and identify the trends that will affect your site in the future.*

Site analysis is an iterative process, done as a series of investigations of your site. Initial data provide a basis for the proposed goals for your project. Continued investigations provide refined data that further clarify project goals and help to set project objectives. Understanding past, present, and future influences in the watershed or region in which your site is located will help you design a project that can adjust to surrounding influences.

4. *Clearly express your project goals and objectives in writing after stakeholders have reached a consensus on their content.*

Goals and objectives help you define your project in terms of desired outcomes and form a basis for common stakeholder expectations. Formal stakeholder agreement and informal stakeholder buy-in to project goals, objectives, and project requirements prevent misunderstandings as your project is implemented, cared for, monitored, and evaluated. This process provides the foundation on which all subsequent decisions are based. Sign off from regulatory agencies at this stage can reduce the likelihood of requirement changes at the time of permit issuance, especially if agency personnel change in the interim.

5. *Identify and eliminate the causes of disturbances that have impaired the ecosystem(s) on your project site.*

Unless you eliminate, or at least minimize, the perturbations that led to the degradation of your site, your restored ecosystem will be subject to the same adverse influences. In cases where the stressors affecting your site are coming from outside your project boundary, you may have to address issues in the region from your project location. The use of a conceptual analytic tool known as SWOT-C is an effective means to conduct an analysis of your site. This analysis allows you to incorporate the site conditions and evaluate them against your initial goals, possibly modifying the goals in response to this information. This tool can also prompt you to consider new goals by discovering opportunities previously not expressed.

6. *Create a design for your project that takes into account not only the project goals and objectives but also the site conditions, adjacent environments, the position of your site in the landscape, and any anticipated effects of climate change.*

Translating your project goals and objectives into a design for your project site is no simple matter. Although your reference model will serve as a guide for what you want to achieve, you will need to determine the required actions on your site to restore missing ecosystem functions and values. You may need to alter the landform to restore the processes that drive the ecosystem, such as surface water patterns or distance to groundwater. This will sometimes require extensive construction activity. At other times, you may only need to prepare a design for the installation of vegetation on appropriate land surfaces. The layout of the plant associations is important, because you will be trying to mimic the natural occurrence of these plant communities with respect to the characteristics of your site. Moreover, your design will attempt to meet the habitat requirements for your target wildlife species. In other cases, you may only need to employ management activities to reintroduce processes such as fire, but you will still face decisions as to what and where to burn. At times, you will need to use a combination of construction and management activities.

In all instances, you will want to integrate your design with the natural and human-made features of the surrounding landscape. Finally, in this day and age of climate change, your design should take into account how your site can provide high ecological integrity and connectivity to be resilient in the face of variable weather patterns, extreme events, and changes in natural processes, such as flood, fire, and pest outbreaks.

7. *Decide on a plan for achieving restoration, and put your plan down on paper (or in digital files) in the form of project documents.*

Project documents are essential for communicating the work to be performed, especially when project implementation will be carried out either by contractors and subcontractors or by different groups of volunteers who have not been involved with the planning and design of the restoration project. Depending on the size, complexity, and nature of your project, your project plans may consist of a single, simple plan or, more likely, as a series of multiple sheets of plan drawings covering different elements of your project, including specifications. Preparing a bill of materials will serve many functions; it is a budgeting tool, an analytical tool, and a communication tool. Finally, your specifications should focus on functionality and include criteria for measuring performance. Together, these project documents are referred to as the bid package or the contract documents.

8. *Conduct detailed planning for the implementation of your project, allowing sufficient lead time for contracting, purchasing, and other tasks well in advance of your construction and installation dates.*

After you have a plan for your restoration project, there are generally a lot of details that must be worked out prior to project implementation. Scheduling and construction planning help to ensure the delivery of materials and equipment when they are needed for installation. A lot of coordination is involved for materials, equipment, and labor to be available on-site at the proper time to maintain the project schedule. Contracting out any project work will require sufficient time for contractor selection. Even if you will be doing all, or most, of the work yourself, you will need to schedule each aspect of the project work and procure all the necessary materials, services, and equipment for getting the job done orderly and efficiently.

9.   *Perform all planning for the procurement of plant materials in advance of introducing plants, seeds, and so forth to your project site.*

Decisions regarding the plant materials involve species selection, propagule type, and quantities. Each decision will affect your project budget and will require careful analysis of plant/propagule availability, labor requirements, timing, and the need for other support materials. In many cases, a long lead time (a year or more) is required for collecting or propagating native plant materials. Even longer times may be needed for locally collected propagules that will be contract grown at a native plant nursery. Prior to planting, consider all of the support systems and stressors that typically are occurring in the landscape.

10.   *Monitor every step in the implementation process, and serve as a coach for everyone performing the management, construction, and installation activities.*

The probability of completing a successful restoration project increases greatly when someone who has been involved with the planning and design of the restoration project is also in a position to oversee the project's implementation. There is no substitute for "eyes on the ground." When implementing management strategies or construction and installation strategies, site inspection by a qualified ecological restoration practitioner is essential for accomplishing a successful project. As site inspector, your role typically will be more like a teacher than a policeman. Your guidance during plant installation will help to ensure that plant materials are handled and installed properly to prevent losses while the project matures.

11.   *Document all work performed, and keep accurate detailed records.*

The plans and specifications for any project are subject to change. Maintaining a record of these changes contributes to the as-built plans. Record any changes in the management treatment regimes or protocols. As you proceed with project aftercare, you can keep detailed records of all maintenance and stewardship activities. Of course, you will also keep all monitoring data, analyses, and evaluation reports. This documentation will serve several functions, including as an aid in understanding what worked and what did not during each step of the project, as a reference for similar future projects, and as a valuable resource for researchers conducting future evaluations of project success.

12.   *Reserve adequate time and resources for maintaining your site improvements, including plantings, and become the long-term steward of your project site.*

Very seldom can one walk away from a project site at the end of the implementation phase. Typically, a period of aftercare is required. One may need to repeat management treatments to obtain the desired effect. Maintenance (short-term care) is generally required for site "improvements," such as repairing damage whether from flood, wind, or vandalism. Your site may need to be mowed or weeded, or it may require repairs to various support items, such as plant protection devices, irrigation systems, and fencing. The most critical time for a project is immediately following construction or installation, commonly continuing for one to three years depending on the ecosystem and site conditions. Many project sites also require stewardship (long-term care) to ensure that the restoration is on a trajectory toward the desired future condition. Project managers need to budget for the staffing and resources required for aftercare as well as obtain the required funds.

13.   *Develop, fund, and implement a monitoring program to evaluate whether, and when, you have achieved your project objectives.*

All restoration projects need to have some type of monitoring and evaluation program with adequate funding relative to the importance and complexity of the project. You need not monitor everything on your project site; rather, monitor those elements that will help you determine if the site is maturing in a way that will meet your performance standards. These standards serve as incremental measures of achieving your project objectives, which were derived from your project goals. Knowing to what degree you have met the objectives for your project will inform your stakeholders of your success or enable you to suggest needs for future remediation.

14.   *Use an adaptive management approach to improve each successive stage of your restoration project and refine the planning for future similar projects.*

Using an adaptive management process can improve your understanding of how to plan, design, and execute successful ecological restoration projects. During installation, things may not go as planned. Changes may be needed to correct poorly performing elements of your project. Do not wait until your project is completed to implement needed changes when they have been identified during the installation process. Use your monitoring data to determine whether any remedial actions need to be taken during the aftercare phase of your project. At the end of each project, your project team can perform an evaluation of what worked and what might be done differently on future projects. Adaptive management is a good way to grow restoration expertise and to improve consumer confidence in the profession.

15.   *Retain your creative spirit and remain open to trying new and innovative approaches for achieving successful restoration projects.*

Creativity and innovation are important attributes to have as you approach your project. No two projects are alike, which is why the guidance in this book is presented as a framework that still requires thought during implementation. Most likely, you will be challenged with a lot of "first time" experiences. With the proper use of the tools included in this book, you will be prepared to solve these challenges.

Help Grow Our Profession

We wrote this book for you—whether you are an ecological restoration practitioner (new to the field or not), a motivated volunteer, or a student looking for a career—in the hope that you will be able to learn from our experiences. Now it is your turn to help us grow the profession of ecological restoration. Here are some suggestions on how you can help:

- Stay informed by joining one or more professional societies that promote ecological restoration.
- Become active in the operation of one of these organizations by serving as a board or committee member, a conference organizer or symposium chair, and so forth.
- Present papers and posters on your restoration work at conferences dealing with ecological restoration.
- Conduct field tours of your restoration project sites.
- Submit articles to organization and chapter newsletters describing lessons learned from your project activities.
- Prepare and publish case studies of your restoration projects.
- Use a variety of communications media to educate the public about the benefits of your restoration project work.

Go Forth and Heal the Earth

We wish you success in your future work as you grow in your chosen profession of ecological restoration. Remember, although you are restoring one project site at a time, you are contributing to the bigger picture of restoring our world's degraded, damaged, and destroyed ecosystems for the purpose of restoring ecosystem health and biodiversity, restoring ecosystem goods and services, creating sustainable livelihoods, and mitigating the effects of climate change.

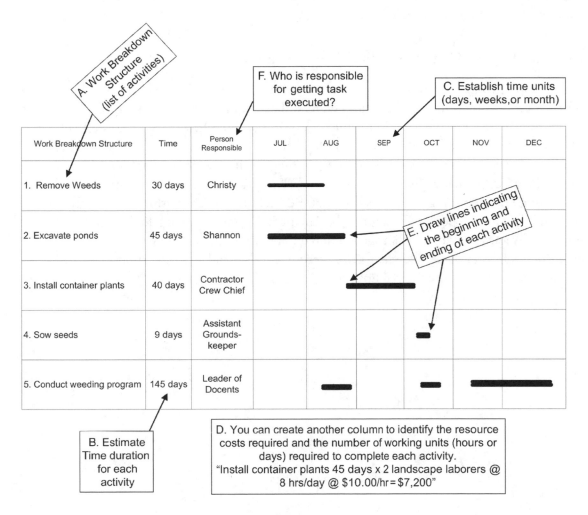

A1-1. A quick primer on developing a Gantt chart diagram.

The numbered items are discussed below.

A. The Work Breakdown Structure column lists a set of discrete actions needed to do the project.

B. Estimate the time duration for each activity—that is, the number of days or hours each activity will take.

C. Establish time units, such as days, weeks, or months. This is a graphic representation of item B showing start and end dates for each activity.

D. Create a column to identify resource needs for the task, such resources as labor, equipment, or money needed for a task.

E. Draw lines indicating the beginning and ending of each activity. Organize these time duration bars in order from start to finish based on any dependent relationships with other activities.

F. Identify by name or position a single party responsible for executing each task. To eliminate miscommunication, avoid shared roles.

How a Gantt Chart Can Help with Scheduling a Project

In this example, you are planning this project to be completed in time for the local fair, which opens October 1. After building your Gantt chart and examining the various entries, you discover an issue: the project finishes late. Referring to figure A1-1, note that the task "Install container plants" has a duration of forty days. You want to begin the activity on August 1 and complete the work on September 15. However, before tree planting can begin, site preparation work must be completed—tasks #1 and #2, "Remove weeds" and "Excavate ponds." The weed-removal task finishes in time, but excavating the ponds does not. Restrictions prevent starting before July 8, causing the end date for the pond excavation to be August 21. With seed sowing to do, planting cannot begin before August 21. This conflict of dates and durations requires problem solving and decisions to meet the October 1 date. The chart illustrates the conflict and assists in choosing which tasks can be managed to meet the desired finish date.

APPENDIX 2. PROJECT COST ESTIMATE WORKSHEET

Project:
Date:

Project Phase	Tasks	Labor	Equipment	Material	Total
PROJECT INITIATION	Project management				
	Develop goals and objectives				
	Project description				
	Candidate site selection				
	Initial cost estimate				
	Needs assessment				

Project Phase	Tasks	Labor	Equipment	Material	Total
ANALYSIS	Site analysis				
	Aerial photography				
	Soil sampling and analysis				
	Well survey and/or installation				
	Land use analysis				
	Hydrologic analysis				
	Water quality				
	Identify degradation sources				
	Assess degradation extent				
	Hazardous waste investigation				
	Groundwater monitoring				
	Resources surveys: biology, geology				
	Topographic analysis				

Project Phase	Tasks	Labor	Equipment	Material	Total
DESIGN	Concept plan				
	Plant palette selection				
	Species assemblages and pattern				
	Vegetative clearing/trash removal				
	Exotic plant eradication				

Project Phase	Tasks	Labor	Equipment	Material	Total
IMPLEMEN-TATION	Initial grading areas and quantities				
	Seed application method				
	Topsoil handling and quantities				
	Salvage and storage plant, rock, logs				
	Soil preparation method				
	Erosion control material				
	Erosion control installation				
	Biotechnical applications				
	Drainage devices and installation				
	Placement of topsoil				
	Fencing				
	Plant protection				
	Soil amendment: chemical				
	Soil amendment: mulch				
	Import topsoil and or fill				
	Confirm ownership of property				
	Resolve any rights to use of land				
	Finalize plant species list				
	Determine plant propagule types				
	Calculate habitat plan areas				
	Calculate plant material quantities				
	Verify availability of species list				
	Order species needing advance time				
	If collecting seed, arrange for preparation and storage				
	Arrange for contract collection and/or growing of seed and plants				
	Transporting plant material to site				
	Supplemental irrigation design				
	Installation and cost of irrigation material				

If growing nursery stock, determine greenhouse needs and storage area and maintenance until planting				
Seed application				
Hydroseeding supplies				
Purchase of soil amendments				
Installation of soil amendments				
Purchase of mulch				
Installation of mulch				
Planting of container stock				
Temporary water devices				
Flagging and construction barriers				
Permanent fencing/barriers, etc.				
Site preparation				
Auger/ditching service				
Inspection services: plant storage				
Inspection services: installation				
Final plans				
Grading plans				
Planting plans				
Irrigation plans				
Specifications				
Development of time estimate				
Development of budget				
Application for permit required				
Hearings, meetings, coordination				

Project Phase	Tasks	Labor	Equipment	Material	Total
AFTERCARE	Supplemental water				
	Cost of water				
	System inspection and repair				
	Operation of system, manual				
	Weed control				
	Herbicide treatments				

	Pest control: insect and others				
	Plant protection maintenance and removal				
	Replacement planting				
	Removal of dead plantings				
	Data recording of maintenance				
	Erosion inspection and repair				
	Site inspection				
	Fencing maintenance, relocation, and repair				
	Litter and debris removal				
	Data retrieval				
	Monitor plant growth				
	Survival				
	Diversity				
	Health				
	Reproductive activity				
	Fire, controlled burns				
	Grazing rotation and durations				
	Fencing relocation				
	Survey animals				
	Birds				
	Mammals				
	Reptiles				
	Amphibians				
	Fish				
	Insect				
	Write up report for surveys				
	Photodocumentation activities				
	Develop maintenance manual for others				
	Training of maintenance staff				

APPENDIX 3. RISK MANAGEMENT STEPS

Identify all known risks (item or task) that could likely occur. Ask: "What can go wrong?"

Assess the probability, or the likelihood, of that item occurring sometime during your project. We use the simple H–High, M–Moderate, and L–Low categories. Enter this rating in column 2 of the risk management worksheet (fig. A3-1).

Risk Management Worksheet

Scope, Schedule, Cost Task or Item	Probability (H,M,L)	Impact (H,M,L)	Exposure Col 2 and Col 3 (HH,MH etc)	Response Measure or Action

A3-1. Risk management worksheet.

Determine what impact the item will have on the project should it occur. Use H, M, or L. Combine the two ratings you have reached to develop a "Total Risk" rating. It is important to keep the order of each rating because they have different meanings. The order of ratings highest to lowest is HH, MH, and HM. The top three ratings generally are all you will have time to analyze and develop an action to implement. A general rule is to handle between eight and twelve of the top issues at a time.

For each item in the top three ratings, develop an appropriate action that will correct the event should it occur. Ask: "How can this be prevented or handled when the problem occurs?" Assign a person to monitor and manage each risk event. The risk management plan allows you to implement immediate responses when risk issues become realities. It can also help reduce or avoid crisis management and may permit you to maintain the project as originally scheduled, despite setbacks or problems that would normally result in delays while the team scrambles for a solution to the problem.

Appendix 4. Project Evaluation and Review Technique

The Project Evaluation and Review Technique (PERT) is a technique for estimating time duration. At times, you will need to estimate work on tasks that have a high number of variables and for which the time required cannot be routinely estimated from a chart. In these cases where a number of variables may influence the duration, using the PERT is a convenient approach. The key is to interview knowledgeable people who are familiar with the task. Ask them a series of three questions:

1. What will be the most likely amount of time in days that it will take to do this task as you understand the project now? (most likely)
2. What is the least amount of time it will take? (optimistic)
3. What is the longest it will take? (pessimistic)

The rule of thumb is that the pessimistic duration should be not more than three times the most likely duration. If this situation does occur, you should obtain more information so that the estimates come closer together.

$$\text{Estimate (duration)} = \frac{\text{Optimistic} + (4 \times \text{Most Likely}) + \text{Pessimistic}}{6}$$

Terms Defined:

Optimistic value is the time it would take to do the task or project without setbacks, delays, or interruption.

Most likely value is the time when personal experience is included and you know what typically will occur during the period of the task or project.

Pessimistic value incorporates those events that can cause delay, such as weather, delivery problems, shortage of materials, labor shortages, and so forth.

APPENDIX 5. SITE ANALYSIS CHECKLIST

	SWOT-C	OBSERVATIONS
GENERAL FACTORS		
Ownership of candidate site		
Easements, prior rights, other conditions		
Historical context		
Prior land use		
Unique site features, structures, landforms		
Current land use		
Political considerations		
Access/access control/human use patterns		
Cultural resources (historical, archaeological)		
Agricultural or other quarantines		
Hazardous waste, debris, etc.		
Stressors on existing ecosystem		
PHYSICAL FACTORS		
Topography		
Slope and aspect		
Elevation		
Geology		
Soils		
Soil chemistry and nutrient status issues		
Topsoil profile and subsoil profile		
Hydrology		
Groundwater status		

Surface runoff		
Water quality status		
Landscape ecology consideration for movement of animals, pollinators, etc.		
BIOLOGICAL FACTORS		
Existing vegetation communities		
Vegetation dynamics of existing communities		
Assess degree of degradation		
Invasive nonnative species		
Habitat value and features		
Wildlife resources		
Endangered, threatened, species at risk, sensitive species presence, or habitat of seasonal use		
ANTICIPATED SITE IMPROVEMENTS OR CONSIDERATIONS		
Grading		
Soil import/export		
Water features		
Drainage/flood control		
Irrigation system		
Buffer zone issues or requirements		
Access control and access to site		
Vandalism issues, need for control features		
Defined candidate area for work		
ADDITIONAL TOPICS:		

Appendix 6. Seed Quantity and Cost Calculation

Figure A6-1 depicts a spreadsheet for calculating the quantity of seed and its cost. Because seed comes in bulk form, it is important to know the quantity of pure live seed in the bulk so that the amount specified for a project is applied. The figure is arranged so it can be used with standard computer spreadsheet software. The example illustrates the different calculations needed to derive a final bulk seed quantity and cost. These calculations are valid only for a particular seed lot with an identified P and G value provided by the supplier, grower, or testing lab. When ordering seed, verify that it comes from one or more lots and determine their respective P and G values.

A	B	C	D	E	F	G	H	I	J	K	L	M
No.	Species	Cost/ Bulk Weight	%P	%G	%PLS	PLS Cost	Design PLS/Unit Area	Bulk Quantity/ Unit Area	Cost/Unit Area	Total Area Units	Total Bulk Weight	Total Cost
					D x E	C/F		H / F	CxI or GxH		I x K	C x L
1	Species A	$4.75	0.6513	0.6592	42.93%	$11.06	5	11.65 lbs	$55.32	1.5	17.47 lbs	$82.98
2	Species B	$3.25	0.3258	0.1525	4.97%	$65.41	4	80.51	$261.65	2.5	201.27	$654.13
3	Species C	$15.50	0.8573	0.7562	64.83%	$23.91	1	1.54	$23.91	4	6.17	$95.64
4	Species D	$4.00	0.5598	0.2599	14.55%	$27.49	3	20.62	$82.48	2.5	51.55	$206.20
5	Species E	$6.35	0.7584	0.6573	49.85%	$12.74	2	4.01	$25.48	0.5	2.01	$12.74
	Total Cost of Seed Order											$1051.68

Seed Quantity and Cost Calculations

Notes:
C,D,E and frequently F are provided by seed supplier or calculated by testing seed
H,K are determined by your design and knowledge of species or advice of seed supplier/grower

A6-1. Seed quantity and cost calculation form.

Column Notes:

A— Number each species to facilitate communication with others and reduce confusion. You can consider this a built-in "plan B," especially if you are ordering several species of the same genus.

B— Use scientific names. There is no substitute for this, especially when dealing with subspecies or variants.

C— This is the cost of bulk seed per unit weight.

D— Purity percentage is a measurement of viable seed and other material in bulk seed. Use four digits (for E also) because this is critical for deriving accurate figures at the end of the calculations.

E— Germination rate reflects the percentage of seed that can germinate. Not all seed is alive, and this factor eliminates those seeds in the final calculations.

F— The product of D and E, which reflects the amount of live seed in a bulk pound of seed.

G— PLS (pure live seed) cost/unit weight is derived using the cost per bulk pound divided by the PLS. In the example, Species B costs only $3.25 per pound and seems like a deal, but when you calculate out the nonseed and dead seed elements, the cost of the seed you want comes to $65.41 per pound!

H— This is the amount of live seed you have decided to put on your site. It is typically given as weight per unit area.

I— Bulk quantity is the calculated amount of units required to apply to a unit area.

J— Cost of seed for the unit area of application. Unit cost will be used to determine the cost of seed for the project or application area.

K— Total area to be seeded by this species. In this example, the areas are different, indicating individual application and not a mix.

L— Total bulk seed for design area to be applied. This is the final quantity required to be applied at the amount identified in column H and is the amount to be ordered from suppliers. In many cases, suppliers can provide you with the quantity by just simply ordering PLS units.

M— Total cost for each species in the seed material list. The sum is the seed amount for this project. Labor and other expenses will need to be entered into the project cost estimate worksheet (appendix 3).

Appendix 7. Plant and Planting Specifications Worksheet

A restoration project has the potential to use a wide array of plant material, from small to large, from seed to mature plant. A plant summary sheet is used to track several aspects of the plant material, including quantities, special soil amendments, type of planting holes, patterns, and spacing. Typically, seed material is placed on a separate sheet, but it can be included here if desired.

Figure A7-1 shows a typical plant sheet that we have used on numerous projects. It clearly identifies important data, from the quantities and sizes of each species to unusual items unique to the project. Soil amendments can be diverse; thus, having clearly identified what is needed will allow for quick calculations in preparing a construction bid estimate. Whether you have your own nursery or intend to purchase plants from a commercial nursery, this sheet will facilitate your plant acquisition. If you cannot confirm the availability of species, you may want to keep a list of acceptable substitutes in your file, in case the contractor cannot locate species on the plant list.

Plant List and Planting Specifications

A plant table has been provided. This table can be modified to meet your needs and conditions.

The summary table is a valuable tool to use in the overall scheduling and installation of the site. At a glance, you can determine the plant quantities, material type, and sizes you need. More importantly, you can compare the list against the material delivered to your site or the material collected by volunteers or contract laborers. This table also serves as a valuable resource when calculating estimates for plant cost and labor needs. Following are some comments on the table by column.

Plant Number—The first column is a useful way to keep track of the different material. Different sizes of the same plant species receive a different number. Doing this with the plant number will permit you to clearly communicate to others even if they do not know plants.

Symbol—Symbols are of your own choosing. They may be a single symbol, as shown in our example, or a pattern, for where larger areas are being planted and it is not helpful to have ten thousand planting symbols of the same plant material in a hectare!

Botanical Name—On the botanical name, you may not want to use the technique shown. These abbreviations are unique to this set of plans and would not necessarily apply to other plans. Abbreviations used by landscape architects in North America are another useful option for clearly labeling plans without cluttering the plan with fully written out names. There may not always be a common name, so inserting "No Common Name" or "NCN" is acceptable.

Size—Any number of choices is available. Using a simple code of numbers that remains constant from project to project, unlike abbreviations, simplifies this table. Measurements may include the size of the container across the top; the shape and length of tubes frequently are given as total length. For larger sizes, "1 gallon," "5 gallon," and so forth are used in the United States. For transplant specimens, the size is the width of the box. This category also includes any material that may be unusual, such as cuttings, rooted cuttings, bare root, or pads (in the case of cacti).

Plant and Planting Specifications

Plant No.	Symbol	Botanical Name	Common Name	Size	Quantity Each	Hole Size		Basin Type	Commercial Fertilizer		Mulch	Staking	Planting On Center Inches	Remarks
						Diameter Inches	Depth Inches		Planting	Plt Estb				
1	@	Cercis occidentalis	Western Redbud	5	84	18	36	-	1 Tab	.5lb	4	1, 2	60	Shrub
2	%	Prunus llicifolia	Holl-leaf Cherry	5	67	18	36	1	-	-	4	1, 2	120	Shrub
3	*	Rhamnus californica	Coffeeberry	1	171	12	24	1	1 Tab	.5lb	4	-	84	Shrub
4	+	Quercus kelloggii	California Black Oak	1	184	12	24	-	-	-	3,4	-	198	Tree
5	^	Quercus wislizenii	Interior Live Oak	1	127	12	24	-	-	-	3,4	-	154	Tree
6	#	Populus fremontii	Fremont's Cottonwood	15	45	24	36	1	-	-	-	1, 2	216	Tree

Note: Underlined portions of botanical name indicate abbreviations used on Planting Plans

Abbreviations
Amend---Amendment
Dia -------Diameter
Ea --------Each
Oz---------Ounces
In -----------Inches
Yd ---------Yard
Lbs --------Pounds
Max -------Maximum
Min --------Minimum

Plt Estb –Plant Establishment
Tab -------Tablet(s)

1 –See Detail X
2 –See Special Provisions
3 –15.0 Feet Diameter Area
4 –In Mulched Areas

Calculated/ Designed by		Date	Revised By	
Checked by			Date Revised	

A7-1. Plant and planting specification worksheet.

Hole Size—Measurement is typically related to the type of material being planted. There are special situations, such as poor soil or the need to reach a groundwater level, that may deviate from the typical planting hole.

Basin Type—Typically in a restoration project, the basin is not used because this would either prevent water from accessing the root ball or keep water close to the plant. This needs to be evaluated with the circumstances taken into account. Planting on slopes will require some type of containment to prevent erosion of the root ball.

Commercial Fertilizer—In most situations, this will not be necessary; however, where needed, this clearly identifies which plants and what quantities.

Mulch—Mulching is a common method to control moisture conditions, provide a limited control to weeds, and provide nutrients into the soil over an extended time. The specifications should describe the type of mulch and the dimensions of application, such as area of coverage or a distance from the plant stem and thickness of application.

Staking—This is sometimes required on restoration projects in unusual conditions. Plants raised for restoration should never be staked from the onset of propagation. When staking is required, it is typically done on larger-sized plants. If your site is subjected to extreme environmental conditions, such as constant or sporadic strong winds or similar weather conditions, then staking would be a logical interim method to get plants established. A specification should show exactly how you want the plant, with the sizes of ties and arrangements.

Planting on Center Inches—The distance between plants is another way to check your numbers and to gauge how dense some plantings will be after they are installed. Generally, the distance is an approximation. One thing to remember is the difference between planting distance between any other plant and the distance between the same species. Trees, for instance, can be spaced anywhere from ten to thirty feet; however, shrubs and herbaceous plants will more than likely be planted between these two trees.

Remarks—This column is simply a way of communicating the type of growth form you want. If there is a possibility of confusing plants or forms, this is the place to make a note. Coastal forms of the same species have much different characteristics, as do plants growing on unique soil formations. The plant taxonomy may not have caught up with what we as restoration designers see and need to maintain.

This plant list table is an extremely effective tool for use in your project. However, it is only as good as the information that is entered. You may want to adapt it to meet your project or style of working. The important point is to have some place where you can obtain similar information easily. Whether you are buying plants from a nursery or propagating them yourself, you will still need to know how many plants of each species and size to produce. Cuttings are another matter and require coordination, so cuttings can be collected at the appropriate time of year. This may be a note for the Remarks column of this table.

When going through the process of developing a restoration design or program, it is always a good idea to review the work. The following checklist has been compiled from the authors' projects or experiences of colleagues. It is designed not as a check-off list but as a prompter or springboard. Although a specific question may not apply to your project, it may cause you to think of something that is related to your project. We encourage you to talk about your project with others; additional points of view can only improve the final outcome.

Project, General

1. Have goals been identified for the project?
2. Are objectives clearly stated for each goal?
3. Are areas identified numerically by vegetation type or habitat type?
4. Are areas and time frames established for different land management techniques, such as controlled fires, grazing, trampling, flooding, and so forth?
5. Have performance criteria been developed with quantitative measurements or clearly understandable qualitative descriptions?
6. Have construction or nonconstruction dates been established to avoid impacting sensitive noise receptors (e.g., residential land uses, nesting)?
7. Are elevations clearly identified?

Site Issues

1. Are there sensitive species in or near the project site?
2. Have sensitive areas, construction, utility and other easements, storage, and staging areas been identified on plans?
3. Have adjacent land uses or potential future land uses been identified from zoning or general plans and incorporated into the design?
4. Is a buffer zone planned for areas adjacent to potentially conflicting land uses or environmentally sensitive areas?
5. Are areas of sensitive resources clearly identified?
6. Are limits of work clearly marked?
7. Are signs, flagging, and fences to be installed in no-work areas?
8. Are work roads, access roads, haul roads, and storage areas identified and clearly marked?
9. Are there provisions for construction-related activities to be identified by the contractor and cleared by appropriate personnel prior to work?

10. Are physical and biological erosion control measures in place?
11. Is solar orientation of the site important for selected species?

Site Preparation

1. Is grading a part of the site preparation?
2. Will tidal channels or creeks be excavated?
3. Is the disposal area of sufficient size to receive the quantity of earth excavated?
4. Will excavation work occur in the vicinity of sensitive areas (e.g., sensitive habitat, endangered species, cultural resources, or any other unique resource)?
5. Are provisions included to control the work in sensitive areas?
6. Is there an erosion control feature for the disposal area?
7. Will the disposal area impact adjacent waters as a result of erosion or sedimentation?
8. Will a portion of the excavated material be placed as a topsoil following final grading?
9. Is a specific topsoil storage area identified on the plans? If not, are conditions described to prevent damaging salvaged topsoil resulting from contamination and erosion.
10. Will soil decompaction be performed before planting, to assist in plant establishment?
11. Will the soil require amending?
12. If mulching, how will mulch remain on the surface during flooding or heavy rainfall?
13. Does the design rely on the use of the topsoil or duff as a source for seed and bulbs, mycorrhizae, and mulch?
14. Will the top layer be mixed with recently exposed parent material to form a mixture of soils?
15. Will the material used for the top layer improve soil texture and provide nutrients?
16. How will irregular surfaces be handled to prevent ponding in unacceptable locations?
17. Will the final grade elevation flood on a regular basis, or will it rely on infrequent storm events?
18. How will the soil be handled to prevent significant loss during flooding?
19. For use as a seed source, do instructions identify a specific method of storing the topsoil, duff, or imported soil and time duration?
20. Are there extensive slopes above graded areas with water runoff features to control erosion onto the site?
21. Do structures have a maintenance obligation?

Vegetation

1. Is the planting density reasonable when compared to the objectives and success criteria?
2. If planting seems thin, are the species involved vigorous growers or quick to reproduce on-site or is recruitment from adjacent natural areas expected?
3. Do the objectives include reference to plant cover, height, or similar structural diversity?
4. Do plant size quantities reflect the demands of the success criteria or time to meet criteria?

5. Are the planting locations reflective of the natural elevations and locations for each species?
6. Have clear and concise planting instructions for each species or group, plant stock type, and size been provided?
7. If planting holes are dug by machinery, is glazing of the hole an issue? (Roughen sides prior to planting.)
8. If cuttings are used, has a source for collecting been identified? Are controls in place? Will the plans ensure implementation at the correct time of year? Are provisions included that allow changing to container plants if the time requirements cannot be met? Are substitute species acceptable?
9. Have provisions been included to adequately address the change in cost when changing plant size or types?
10. Have provisions identified all the necessary biological material, such as seed, mycorrhizae, and so forth?
11. Is there a requirement to the contractor to provide proof of retaining plant material in the quantity and sizes specified by the plans at the beginning of project? If not, what other provisions have been made to ensure the availability of plant materials at the appropriate stage of site construction?
12. Is site-specific plant material required? Has scheduling included sufficient and practical time to collect or grow in advance to maintain the project completion schedule?
13. Are all species included in the design available either commercially or by contract collector or grower and at the quantities specified?
14. Do any species presently occur on an official endangered or threatened list? If any species do occur on the list, have appropriate permits been obtained? Or is the contractor or installer required to obtain the appropriate permits?
15. Is there an inspection requirement with a rejection basis for any material, plant, or work?
16. Do seed mixes include species with both common and scientific names, total pound per acre, germination, and purity levels?
17. Are there inspection methods for checking seed germination and purity rates?
18. Is there a provision for the project inspector, project manager, or other responsible person to retain a small sampling of the seed before application?
19. Are re-application methods described?
20. Is a provision included addressing alternative seeding time if unable to apply during optimum time?
21. Is there a specification that controls the duration time between planting and first watering?
22. If staking is required, has a detailed program been discussed outlining the process of checking straps to prevent damage to plants and eliminating the stakes during the maintenance period?

Water

1. Have the flood stage and water table depths been investigated?
2. Is the final elevation of the project site low enough to permit consistent flow through the site?

3. Will existing structures be eliminated or modified to permit water to enter the project site?
4. Is supplemental irrigation required for the project?
5. Has the water source to be used for irrigation or flooding been described?
6. Is irrigation or flooding included in the plan?
7. Is a consistent and predictable water source with good water quality being used for the site?
8. Will all areas of the restoration project have equal requirements for supplemental watering? (If not, the irrigation system should be designed for independent operation in each of the specific locations.)
9. If overhead impact sprinklers are used, is there sufficient overlap designed to offset the obstruction from the planting? Are sprinklers elevated higher than normal to offset the obstruction from seedlings, plants, and weeds?
10. If installed aboveground, have measures been included to ensure reliable operation?
11. Have flow and pressure calculations been made for irrigation system demand?
12. Are drip irrigation emitters sufficiently large to prevent clogging by algae, mineral deposition, and insect activity?
13. Is water accessing the site through a structure that will not impede the flow rate?
14. Have appropriate features been included with culverts to prevent scouring?
15. Will rock slope protection be used? Is there sufficient clearance to prevent clogging or impeding water flow?
16. If there is more than one water feature included in the design, what controls the water level and will the features act independently?
17. For water features designed for foraging habitat, have water depths been calculated to meet the requirements of targeted animal groups?
18. Will the culverts have screens that preclude fish moving into the project area?
19. Will a "natural opening" be created within the vegetation? How will it be kept open? Is a maintenance schedule included?
20. If water is introduced into the project by a pump, is it sized correctly to provide adequate volume to compensate for percolation, evaporation, and transpiration during the hottest period of the year?
21. If using a siphon inlet, what measures have been included to prevent vandalism?
22. If a tidal gate is used, will algae mats and flotsam be controlled to prevent gate malfunction?
23. If water source is from a water utility company, have capacity fees been included in the estimate?

Maintenance

1. Do specifications address weeding maintenance? Is there a definition of what constitutes a weed, and the maximum size, age, or coverage? Is there a removal schedule?
2. Is there a provision for maintaining protection devices after they are installed?
3. Is there provision for removal of stakes, protections screens, plant sleeves, and so forth when no longer required?

4. If fertilizer is being applied topically, are provisions made to control weed growth, if surface applied?
5. Is there a requirement to install fencing or other barriers, including plant protection screens and so forth?
6. Will maintenance of the project be absorbed into an existing maintenance department budget? If not, is separate funding allocated?
7. Are domestic ducks present in the location? (Consider them predators on new plantings.)
8. What provisions have been included to handle possible vandalism of plantings, fencing, signage, and irrigation equipment?
9. Is replacement planting included in the maintenance contract?

Operations and Programs

1. Is a supply of water accounted for through agreement or water rights?
2. Has a budget been developed for controlled burning operations?
3. Has the frequency of burning been included in a program budget?
4. Has coordination been conducted with any agency or department having jurisdiction over burning in your area?
5. Have appropriate buffer or fuel lines been planned to contain a burn?
6. Is there a constant supply of cattle and horses for grazing operations?
7. Are there provisions for keeping herd numbers stable? Are there established introduction and removal dates?
8. Will additional sources of feed be required for a maintained herd?
9. If the herd is from another source, have agreements, contracts, or other arrangements been finalized, with introduction and removal dates?
10. Are introduction and removal dates for maintained herds based on calendar or vegetation removal effect?
11. If the goal is trampling or soil disturbance by cattle, has testing been conducted to know the number and duration of cattle visiting on-site?
12. For projects emphasizing weed removal and natural recruitment, has a schedule been developed for monitoring?
13. Are provisions made for replacement of any equipment needed on-site?
14. If using herbicides, have necessary application permits been obtained? Is more than one person holding a permit?
15. Have arrangements been made for disposal or processing of weed biomass off-site or on-site?
16. Have life histories of the weeds involved been identified (i.e., germination, seed set)?
17. What post–weed removal maintenance will be required to ensure native plant reestablishment?
18. If burning removed weeds is a part of the plan, is an air quality permit required? Are there date or seasonal restrictions?

Special Considerations

1. Will there be pest removal or animal control programs?
2. If so, have appropriate agencies been contacted and permits obtained?
3. Have approved equipment, traps, bait, and other devices been included in the budget?
4. How will you address flotsam from watercourses? Will there be a range of tolerable debris buildup before needing removal?
5. For exclosures, will there be a maintenance program to ensure fence integrity?
6. Has replacement material been included in the budget along with personnel hours to install and maintain?
7. If excavation is near a sensitive area, will resources require noise or dust abatement conditions? Is there some other work restriction that may control the type of equipment being used?
8. Is there a concise list of activities and items that are the responsibility of the contractor during installation, during plant establishment, and during longer time periods?
9. Is it anticipated that beaver, deer, elk, rabbit, or other animals may frequent the site?
10. Are there provisions to protect the site from cattle, horses, wildlife, people, or off-road vehicles during the initial establishment stages?
11. If there is current recreational activity on the property, have restrictions been applied to this activity during the construction and establishment stages?

Appendix 9. Permits, Agreements, and Consultations That May Be Required for Ecological Restoration Projects in the United States

Permit, Agreement, or Consultation	Implementing Agency	Enabling Legislation	Critical Resources or Concerns	Circumstances Typically Required
Federal Government				
Archaeological Survey and/or Excavation Permit	Federal or Tribal Agency managing public/tribal lands	Archaeological Resources Protection Act of 1979 (ARPA)	Protection of archaeological resources	Required to conduct an archaeological survey and/or excavation on federal or tribal lands.
Bird Banding and Marking Permit	U.S. Geological Survey (USGS) Bird Banding Laboratory	Migratory Bird Treaty Act of 1918 as amended	Protection of migratory birds	Required to capture and handle any migratory bird for banding or marking purposes. A state permit may also be required.
Coastal Zone Federal Consistency Review	Lead State Agency in cooperation with National Oceanic and Atmospheric Administration's National Ocean Service Office of Ocean and Coastal Resource Management	Coastal Zone Management Act (CZMA) of 1972	Protection of coastal resources	Federal consistency is the CZMA requirement where federal agency activities (consistency determinations, financing, or other administrative actions) that have reasonably foreseeable effects on any land or water use or natural resource of the coastal zone.
Enhancement of Survival Permit via Candidate Conservation Agreement with Assurances (CCAA)	US Fish and Wildlife Service (USFWS)	Endangered Species Act (ESA) of 1973	Protection of habitat for proposed, listed or candidate endangered or threatened plant or animal species	Landowner voluntarily commits to conservation actions that will help stabilize or restore the species with the goal that listing will become unnecessary. Agreement provides landowners with assurances that their conservation efforts will not result in future regulatory obligations in excess of those they agree to at the time of entering into the agreement.
Enhancement of Survival Permit via Safe Harbor Agreement (SHA)	US Fish and Wildlife Service (USFWS) or National Oceanic and Atmospheric Administration (NOAA)	Endangered Species Act (ESA) of 1973	Federally listed threatened and endangered species	Voluntary agreement with nonfederal landowner that protects private or other nonfederal property owners whose actions contribute to the recovery of a species listed as threatened or endangered under the Endangered Species Act. Participating property owners receive formal assurances that the federal government will not require any additional or different management activities without the property owner's consent.

Permit, Agreement, or Consultation	Implementing Agency	Enabling Legislation	Critical Resources or Concerns	Circumstances Typically Required
Endangered Species Consultation and/or Biological Opinion (USFWS)	US Fish and Wildlife Service (USFWS)	Endangered Species Act (ESA) of 1973	Federally listed species and critical habitat	Required for wildlife habitat restoration projects on any federal lands or that use federal funds, in any areas designated as critical habitat for the recovery of a listed species. Required for stream restoration projects where federally listed freshwater fish species and other listed aquatic species (e.g., amphibians, snails) are known to occur.
Endangered Species Consultation and/or Biological Opinion (NMFS)	National Marine Fisheries Service (NMFS)	Endangered Species Act (ESA) of 1973	Listed species (marine and anadromous salmon) and critical habitat	Required for stream restoration projects that may affect anadromous fish species and critical habitat.
FEMA Letter of Map Revision (LOMR)	Federal Emergency Management Agency (FEMA)	National Flood Insurance Act of 1968, as amended, and the Flood Disaster Protection Act of 1973, as amended	Protection of floodways and prevention of flood hazard	A Letter of Map Revision (LOMR) is generally based on the implementation of physical measures that affect the hydrologic or hydraulic characteristics of a flooding source and thus result in the modification of the existing regulatory floodway. Applies to restoration projects involving dam removal—if the floodplain or flood elevations change as a result of dam removal. May also need to do this to prove that a restoration project is not in a regulated floodplain.
Incidental Take Permit—Section 10(a)(1)(B) Permit	US Fish and Wildlife Service (USFWS)	Endangered Species Act (ESA) of 1973: Section 10(a)(1)(B)	Protection of all federally listed animal species	Incidental Take Permits are associated with Habitat Conservation Plans. Additional reasons for a permit include but are not limited to construction and development and in-stream and watershed activities that may impact listed species.
Incidental Take Permit via Habitat Conservation Plan	US Fish and Wildlife Service (USFWS)	Endangered Species Act (ESA) of 1973: Section 10(a)(1)(B)	Protection of all federally listed animal species	Permits otherwise legal activities that result in the "incidental" taking of a listed species as long as there is an approved Habitat Conservation Plan (HCP).
MMPA Incidental Take Authorization or Letter of Authorization	National Marine Fisheries Service (NMFS)	Marine Mammal Protection Act of 1972	All marine mammals	Restoration project that "takes" or harasses a marine mammal, including the relocation of marine mammals.
Pesticide Use Permit	Multiple agencies	Federal Insecticide, Fungicide, and Rodenticide Act (FIFRA)	Protection of wildlife and water quality	Application of herbicides to control weeds on federal lands.
Research/Recovery Permit (Recovery and Interstate Commerce Permit)—Section 10(a)(1)(A) Permit	US Fish and Wildlife Service (USFWS)	Endangered Species Act (ESA) of 1973: Section 10(a)(1)(A)	Protection of all federally listed species	Scientific research or activities to enhance a listed species propagation or survival that would directly "take" a wildlife species. Activities include abundance surveys, genetic research, relocations, capture and marking, telemetric monitoring, and seed collecting from endangered plant populations.

Permit, Agreement, or Consultation	Implementing Agency	Enabling Legislation	Critical Resources or Concerns	Circumstances Typically Required
Research/Recovery Permit (Recovery and Interstate Commerce Permit)—Section 10(a)(1)(A) Permit	National Marine Fisheries Service (NMFS)	Endangered Species Act (ESA) of 1973: Section 10(a)(1)(A)	Protection of federally listed anadromous fish species	Same as above but addresses specifically anadromous species, including salmon. Examples include fish surveys, genetic research, hatchery operations, relocations, capture and marking, and telemetric monitoring.
USACE General Permit Program—Programmatic General Permit	US Army Corps of Engineers (USACE)	Rivers and Harbors Act of 1899: Section 10, and Clean Water Act of 1972: Section 404, as amended	All navigable waters, tidal waters, wetlands, and waters of the United States	Programmatic General Permits are issued by the USACE on a state-by-state basis.
USACE Nationwide Permit (NWP) Program—Pre-Construction Notification	US Army Corps of Engineers (USACE)	Rivers and Harbors Act of 1899: Section 10, and Clean Water Act of 1972: Section 404, as amended	All navigable waters, tidal waters, wetlands, and waters of the United States	Nationwide Permits related to restoration include NWP 5 (Scientific Measurement Devices), NWP 6 (Survey Activities), NWP 13 (Bank Stabilization), NWP 27 (Aquatic Habitat Restoration, Establishment and Enhancement Activities), NWP 33 (Temporary Construction, Access, and Dewatering), NWP 37 (Emergency Watershed Protection and Rehabilitation), and NWP 43 (Stormwater Management Facilities). USACE District Offices may set additional requirements.
USACE Section 10 Permit	US Army Corps of Engineers (USACE)	Rivers and Harbors Act of 1899: Section 10	All navigable waters and tidal waters	Required for construction of piers, breakwaters, bulkheads, jetties, weirs, and intake structures; dredging or disposal of dredged material; and excavation, filling, or other modifications to waters of the United States. Work in, over, or under navigable waters or that affects course, location, condition, or capacity of such waters.
USACE Section 404 Permit	US Army Corps of Engineers (USACE) (also requires state approval under Section 401)	Clean Water Act of 1972: Section 404, as amended	Wetlands and waters of the United States	Required for projects that involve (1) discharge of fill or dredged material into the waters of the United States, including wetlands; (2) site development fill for residential, commercial, or recreational developments; (3) construction of revetments, groins, breakwaters, levees, dams, dikes, and weirs; and (4) placement of riprap and road fills. Applies to any restoration projects on federal lands or that use federal funds.
State and/or Regional Government				
Archaeological Investigation Permit	State Historic Preservation Office (SHPO)	State Statutes	Archaeological resources located on state lands	Required to conduct archaeological field studies on state lands or within state-controlled waters.
Clean Water Act Section 401 Water Quality Certification and/or Waste Discharge Requirements Determination	State Water Resources Control Board or Water Quality Control Board or State Dept. of Environmental Protection, etc.	Federal Clean Water Act of 1972: Section 401, as amended	Compliance with state water quality standards	Any restoration project that involves dredge or fill activities that may result in a discharge to US surface waters and/or "Waters of the State" is required to obtain a CWA Section 401 Water Quality Certification and/or Waste Discharge Requirements (dredge/fill projects) Determination. Also applies to any restoration project that requires a federal permit.

Permit, Agreement, or Consultation	Implementing Agency	Enabling Legislation	Critical Resources or Concerns	Circumstances Typically Required
Coastal Development Permit	State Coastal Commission or Local Coastal Zone Commission	Federal Coastal Zone Management Act of 1972 and State Coastal Zone Management Acts	Protection of coastal resources	Projects located within a state-designated Coastal Zone, including projects to return a project site to a predevelopment condition. Includes wetland and coastal lagoon restoration projects.
Cultural Resources Consultation	State Historic Preservation Office (SHPO) and/or Tribal Historic Preservation Office (THPO)	National Historic Preservation Act (NHPA) of 1966 and State Historic Preservation Acts	Protection of archaeological and cultural resources	Consultation with SHPO and/or THPO for projects potentially affecting sites of archaeological and cultural resources on state and federal lands (NHPA Section 106).
Historic Area Work Permit or Historic Structure Modification—Project Authorization	State Historic Preservation Office (SHPO) and/or Tribal Historic Preservation Office (THPO)	National Historic Preservation Act (NHPA) of 1966 and State Historic Preservation Acts	Protection of historic structures	Same as above but for historic structures, such as breaching a historic dam.
Hydraulic or Hydrology Project Approval or Permit	State Division or Department of Fish & Wildlife or other State Agency	State Statute	Stream channel within the Ordinary High-Water Mark (OHWM)	Any construction activity that uses, diverts, changes, or obstructs the bed or flow for state waters. Required for salmonid habitat restoration projects in some states, including log, logjam, or debris removal.
Incidental Take Permit	State Division or Department of Fish & Wildlife	State Endangered Species Act	Rare, endangered, and threatened species	Activities likely to impact state-listed species, including habitat enhancements for the species.
Lake or Streambed Alteration Agreement	State Division or Department of Fish & Wildlife	State Fish and Wildlife Code	Lakes, streambeds, and stream banks	Activities that could significantly modify a lake, stream, or river by substantially changing the bed, channel, or bank of a river, stream, or lake. May apply to alteration of riparian vegetation, including the removal of exotic vegetation.
Levee Plantings Notification and/or Approval	Levee District and/or Reclamation District and/or Flood Protection Board	USACE O&M Guidelines and State Flood Management Agency Guidelines	Levee and floodway maintenance	Applies to plant species selection on, or near, flood control levees and the management of plants on or in proximity to a levee.
Mosquito Control Consultation	Mosquito Abatement District	State and Local Laws and Regulations	Mosquito production: mosquito vectors of mosquito-borne diseases	Projects resulting in creating increased larval habitat for mosquitoes. Sometimes applies to stream restoration projects.
National Pollution Discharge Elimination System Permit	Environmental Protection Agency (EPA) regional office and/or cooperating local agencies	Federal Clean Water Act	Protection of surface water quality and prevention of sedimentation	Project sites greater than one acre that expose soil to potential soil erosion. May need a Stormwater Pollution Prevention Plan (SWPPP) to prevent polluted runoff.
Pesticide Use Recommendation Form	State Department of Pesticide Regulation	State Statute or Policy	Protection of fish, wildlife, and aquatic resources	When pesticides and/or herbicides are used.

Permit, Agreement, or Consultation	Implementing Agency	Enabling Legislation	Critical Resources or Concerns	Circumstances Typically Required
Prescribed Burn Permit	State-Regional Air Quality Control Board	State-Regional Air Quality Code and Regulations	Protection of air quality	Use of prescribed fire as a management tool.
Plant Collecting Permit or Plant Research Permit	State Division or Department of Fish & Wildlife	State Endangered Species Act and/or State Native Plant Protection Act	Rare, threatened, and endangered plant species	Projects involving state-listed plant species, including seed collection or other propagules for propagation or recovery actions, transplanting of a threatened or endangered plant species.
Scientific Collecting Permit or Scientific Take Permit	State Division or Department of Fish & Wildlife	State Endangered Species Act and/or State Statute	Protection of wildlife resources and State-listed species	Monitoring programs that involve sampling involving live trapping or other forms of sampling that may injure protected wildlife species. Also for sampling fish populations before, during, and/or after a project.
Timber Harvesting Permit	State Department or Division of Forestry	State Forest Practices Act and Code	Protection of forest resources	Projects that involve removal of merchantable timber (trees); may be required for projects involving tree thinning and/or tree removal.
Trapping Permit or License	State Division or Department of Fish & Wildlife	State Fish and Wildlife Code		Temporary trapping and removal of native wildlife nuisance species (e.g., beaver) that could potentially destroy plantings.
Water Rights Permit or Water Diversion Limited License	State Water Board or State Division of Water Rights	State Water Resources Code	Use of water resources without water rights	Projects needing to divert (including pumping) and use water from a stream on a short-term or fixed duration for irrigating nonriparian land.
Local Government				
Agricultural Land Use Conversion Permit	County, Parish, Township, City or Town	Local Land Use Conversion Ordinance	Preservation of agricultural lands	May be required for conversion of agricultural lands to wildlife habitat. In some states, state has jurisdiction.
Burning Permit	County, Parish, Township, or Town Air Quality Control District and/or Local Fire District	County, City, etc. Ordinance or Agricultural Commissioner Policy	Air quality and fire hazard	Burning to eliminate debris. Burning to control weeds. Prescribed fire.
Encroachment Permit	Flood Protection Board or Flood Control District, or County, Parish, Township, City, or Town	State Stature or Policy	Potential hydraulic impacts of development (including plantings) in floodways	Encroachment into rivers, waterways, floodways, and floodplains. Assessment of project site "improvements" on flood flows and elevations. A flood-neutral planting design may be required.
Erosion Control Permit	County, Parish, Township, City, or Town	County, City, etc. Erosion Control Ordinance	Soil erosion and sedimentation	Any land-disturbing activity in excess of a specified area (square feet), or on a slope greater than a certain percent, or exceeding volume quantity of grading, or along a waterway or shoreland zone.
Grading Permit	County, Parish, Township, City, or Town	County or City Grading Ordinance	Whenever construction involves moving a certain amount of earth	Projects involving channel relocation or reconfiguration. Projects that involve lowering of land elevation relative to water surface elevation (e.g., wetland, riparian).

Permit, Agreement, or Consultation	Implementing Agency	Enabling Legislation	Critical Resources or Concerns	Circumstances Typically Required
Herbicide Use Report	County, Parish, Township, or Town Agricultural Commission	County, etc. Agricultural Commissioner Policy	Documentation of herbicide use	Herbicide spraying to control weeds.
Reclaimed Water Use Permit	County, Parish, Township, City, or Town	County, City, etc. Ordinance or Policy	Protection of water quality	Projects that intend to use treated wastewater for wetlands and possibly for the irrigation of upland plantings.
Stormwater Management Permit	County, Parish, Township, City, or Town	County, City, etc. Stormwater Management Ordinance	Protection of water quality	Activities that may significantly increase runoff, flooding, soil erosion, or water pollution or significantly impact a lake, stream, or wetland area.
Tree Removal Permit	County, Parish, Township, City, or Town	County, City, etc. Tree Protection Ordinance	Protection of designated heritage trees and trees over a certain diameter	Where tree protection ordinances are in effect.
Well Deconstruction or Destruction Permit	County, Parish, Township, City, or Town Environmental Health Agency	State Standards and County, City, etc. Ordinances	Prevention of groundwater contamination	Decommissioning of unused or abandoned wells. Must follow state standards and any additional local requirements.
Well Drilling Permit	County, Parish, Township, City, or Town	County, City, etc. Well Drilling Ordinance	Prevention of groundwater contamination	Drilling of a production well for watering plantings or other purposes.

General Factors	SWOT-C	Comments
Political considerations	S	Removal of tall skyline eucalyptus—visual impacts for park users and nearby residences.
		Consult with landscape architects for visual study.
Historical context	S	Canyon area operated as a ranch beginning in mid-1800s; preserve staff says the canyon was used mainly for cattle grazing; photos in the old adobe show sycamores, oaks, and cottonwoods in early years of ranch operations.
Hazardous waste	—	No evidence of old fuel tanks or washout pads.
		Check county records for any information regarding historic Rancho operations.
Resource constraints	C	Existing native vegetation in area; stream is live.
		Flag off areas; develop bypass system for water.
Historical/Archaeology	C	Grave site near large oak; Spring House (foundation is being undermined by palms); Rancho adobe is on National Registry.
		Coordinate with preserve staff on requirements and restrictions.
Wildlife	S	Great horned owl's nest in eucalyptus grove.
		Preserve and relocate nest; consult with Biology to determine best time or season for relocation.
Human use patterns	T	Trails proliferate throughout project area; daily use includes hikers, joggers, equestrians, mountain bikers; school programs near Rancho.
		Coordinate with preserve staff regarding trail closure during construction and plant establishment.
Identify ecosystem stress points	W	Large stands of eucalyptus trees cover streambed; large eucalyptus grove at headwaters of Peñasquitos Creek; several individual eucalypts scattered throughout canyon.
	W	Volunteer palms (*Phoenix* spp.) dominate the Spring House and creek area.
	W	Large quantity of leaf litter and shallow roots.
Ownership of candidate site	S	Entire site is owned by County.
		Obtain parcel map from County records.

Constraints	C	Will trails need to be kept open during construction and plant establishment?
	C	Heavy trail use in a.m./p.m. periods; may require work windows during installation.
Easements, prior rights	T	Utility corridors crisscross the site; sewer manholes are visible on main service road; water main runs north/south; overhead electrical lines share same easement.
		Request easement restrictions and details from utility companies.
	T	Surface erosion along main service road.
Agricultural quarantines		None.
Land use	S	Project site is within an open space park operated by the County; residential land uses are located at the top of the north and south canyon rims.

Physical Factors		
Define candidate area		Three-acre site within the park; project area is covered with non-native eucalyptus.
Landscape ecology considerations	S	Part of a large riparian vegetation community with direct connection to adjacent hillsides. Wildlife can reach site easily and move up and downstream without any obstacles. Stream flows through vegetation with several channels present. Migration of fish only downstream as waterfall below site prevents upstream movement.
Hydrology	S	Peñasquitos Creek flows year-round; three minor overflow channels run parallel through site and appear to carry water during storm events.
Groundwater	S	Natural spring feeds creek at Spring House.
Surface water	S	Appears clean; is free from debris and sedimentation though urban runoff must contribute to status as year-round creek.
	T	Ranch House trail has potential sedimentation problem.
Water quality	?	Order test for salinity/nitrogen/phosphates/boron/and heavy metals.
Topography	S	Wide canyon bottom; runs east to west relatively flat; mesas located on the north and south of canyon.
Elevation	S	~ 200 feet +/– MSL

Slope and aspect	S	South canyon wall is steep (~30%) and covered with chaparral; north canyon wall is flatter and dominated by nonnative grasses; gets full sun; canyon bottom in full sun.
Soil testing	O	Loamy sand in project area, with sandy and gravelly loam near main service road and trails.
		Determine whether any data are available from County.

Biological Factors	SWOT	Comments
Determine successional patterns of existing vegetation	W	Vegetation currently in "climax" condition with mature and developing eucalyptus trees and lack of understory. Small area on perimeter still exhibiting diversity of native plants but in small clusters. Some areas with young eucalyptus indicating the process of degradation is ongoing. Native vegetation adjacent to eucalyptus colonies shows typical pattern riparian altered regularly by flood flows.
		Establishing willow scrub woodland in area will serve to bring the upstream and downstream vegetation into one continuous cluster, providing enhanced microhabitats.
Identify habitat values and features	O	Tall trees provide roosting and nesting habitat for raptors and larger birds. Owl nesting and several hawks observed roosting in various locations. Small fishes in water from upstream; no access from downstream. Mammals frequent area; bobcat, fox, raccoon, and skunk are common in this area of the park. Signs of deer seen along riparian area on trails.
		Project will provide foraging areas within the current eucalyptus grove. The project will encourage more frequent visitation and increase the diversity of animals using the vegetation to more than the larger predatory birds.
Assess degree of degradation	W	Large areas along creek banks devoid of native understory; eucalyptus dominates.
	W	Soil surface appears to have about 6 to 10 inches of eucalyptus leaves and seeds.
		Talk with team about removing top soil layer to rid the site of eucalyptus seed bank.

Wildlife resources	C	Great horned owl nesting on-site in eucalyptus.
		Move nest into sycamore tree nearby.

Anticipated Site Improvements		
Grading	S	Only minor grading may be necessary to clean up after tree stump removal.
Soil import/export	W	Remove 6 to 10 inches of soil in eucalyptus groves to remove seed bank?
Drainage/flood control	O	Use existing overflow channels to divert stream during construction.
		Consider enhancing stream crossing and reducing sedimentation potential.
Buffer requirements		None.
Access/access control	W	Temporary fencing around restoration areas will be necessary to protect new plantings.
		Provide access gates for plant establishment.
		Coordinate with preserve staff regarding fence locations.
Utility service	O	Electrical service connection at Ranch House; north side of creek approximately 400 feet from site.
Water		Point of connection for city water service located on north side of creek approximately 600 feet away from site.
Electric		Consider solar-powered controller, especially because irrigation will be temporary.
Irrigation		Required for two years.

S–Strength
W–Weakness
O–Opportunity
T–Threat
C–Constraint

actions. Measures that are undertaken to achieve stated objectives.

adaptive management. A structured, iterative process of optimal decision making in the face of uncertainty, with an aim to reducing uncertainty over time via system monitoring.

anthropogenic environmental stressor. A human-induced environmental condition or recurring event that is detrimental to the stability or development of an ecosystem.

best management practices (BMPs). Techniques, processes, activities, or structures used to mitigate direct or indirect impacts.

bioengineering. A branch of engineering in which live plants and plant parts are used as building material for erosion control and landscape restoration in contrast to conventional engineering where only dead materials are used. Also referred to as soil bioengineering.

biotechnical stabilization. The integrated or combined use of living vegetation and inert structural components to stabilize a slope.

buffer zone. Areas between core protected areas and the surrounding landscape or seascape that protect the core area from potentially damaging external influences; they are essentially transitional areas.

community. An assemblage of organisms occurring in a landscape or at a specified location; typically used in combination with a taxonomic group (plant community, insect community, epiphyte community).

compensatory mitigation. An approach or strategy used by government agencies to require that unavoidable environmental damage is compensated by ecological restoration or another activity (rehabilitation, reclamation, enhancement, and so forth).

connectivity. Landscape connectivity can be defined as the degree to which the landscape facilitates or impedes movement of materials between resources patches.

corridors. A narrow strip landscape used by wildlife and potentially allowing movement of biotic factors between two areas.

creation. The intentional replacement of an ecosystem with another kind of ecosystem of alleged greater value, as has commonly been required for satisfying compensatory mitigation requirements.

cultural ecosystems. Ecosystems that have developed under the joint influence of natural processes and human-imposed organization.

design-build contractor. A contractor who prepares restoration project plans for stakeholder approval and then constructs the project under a single contract.

ecological attributes. Biophysical (composition, structure, abiotic/landscape support) and emergent (functionality, complexity, self-organization, resilience, self-sustainability, biosphere support) properties of ecosystems.

ecological engineering. The manipulation and use of living organisms or other materials of biological origin to solve problems that affect people.

ecological integrity. Ecological integrity is the state or condition of an ecosystem that displays the biodiversity characteristic of the reference—such as species composition and community structure—and is fully capable of sustaining normal ecosystem functioning.

ecological restoration. The process of assisting the recovery of an ecosystem that has been degraded, damaged, or destroyed.

ecological restoration practitioner. An individual who is actively engaged in the various phases and aspects of ecological restoration and who is knowledgeable in the concepts of restoration ecology and the principles and practices of ecological restoration.

ecological restoration project. A planned undertaking designed to recover degraded, damaged, or destroyed ecosystems at specific project sites within defined (i.e., mapped) boundaries. Ecological restoration projects attempt to restore most if not all of the attributes of restored ecosystems.

ecological trajectory. The projected developmental pathway of the ecological attributes—biotic and abiotic—of an ecosystem through time.

ecosystem degradation. The incremental and progressive impairment of an ecosystem on account of continuing stress events or punctuated minor disturbances that occur with such a frequency that natural recovery does not have time to occur.

ecosystem processes. The underlying processes of an ecosystem, such as energy transfer, primary production, food chain dynamics, hydrological pathways, and nutrient cycling. Inextricably linked with ecosystem structure but not synonymous with ecosystem functioning.

ecosystem recovery. The rate and manner in which the ecosystem subsequently returns to its unstressed condition or follows a chronological sequence of development (often termed trajectory) that would coincide with an unstressed reference condition, if recovery indeed occurs.

ecotone. A transition zone between ecosystems.

edge effect. The phenomenon of increased wildlife abundance and diversity along the edge between two adjacent plant communities.

electroconductivity. Conductivity of electricity through water or an extract of soil. Commonly used to estimate the soluble salt content in solution. See also **soil electrical conductivity**.

environmental engineering. The integration of science and engineering principles to improve the natural environment (air, water, or land resources); to provide healthy water, air, and land for human habitation (house or home) and for other organisms, and to remediate pollution sites.

environmental stressor. A normally occurring condition or recurring event that is more detrimental to some species than to others and that largely determines species composition and abundance in an ecosystem. Examples of stressors include freezing temperatures, drought, salinity, fire, and unavailability of nutrients.

fabric mulch. Synthetic material placed on the ground around plants to control weed growth. Also referred to as a weed mat.

fabrication. Establishment of an ecosystem on land that previously did not have this ecosystem. Also called creation.

flaming. Eradication of weeds and invasive plants by means of burning plant parts using a propane torch.

function. The dynamic aspects of ecosystems, such as photosynthesis, primary production, sequestering and recycling of mineral nutrients, and maintenance of food webs. Sometimes restricted in meaning to these metabolic activities and sometimes expanded to include all ecosystem processes.

geomorphology. The description and study of landforms.

girdling. The complete removal of a strip of bark (consisting of cork cambium, phloem, cambium, and sometimes going into the xylem) from around the entire circumference of either a branch or trunk of a woody plant. Girdling results in the death of wood tissues above the damage. Also called ring barking or ring-barking.

habitat. The resources and conditions present in an area that produce occupancy—including survival and reproduction—by an organism: habitat is organism specific.

hard seed. Any seed with a tough impervious outer coat that will not allow the entry of water. Germination cannot occur until the seed coat is ruptured, either by scarification or by microbial action.

hardpan. A soil layer with physical characteristics that limit root penetration and restrict water movement.

herbivore. An animal that feeds on plants. *Herbivory* is the state or condition of feeding on plants.

hydroperiod. The duration that a soil or substrate is saturated or inundated over the course of a year or other time period.

imprinter. A roller that makes microcatchments with teeth, cones, or V-shaped ridges arranged in a pattern to direct water flow; typically towed by a tractor, bulldozer, or other heavy equipment.

indigenous. Native to a given location.

indigenous people. A body of persons having originated in and being produced, growing, living, or occurring naturally in a particular region or environment united by a common culture, tradition, and kinship; exhibiting the practice of common social, economic, environmental, and spiritual beliefs.

inoculation. The act of introducing mycorrhizae or bacteria (inoculum) to a plant.

keystone species. A species that has a substantially greater positive influence on other species than would be predicted by its abundance or size.

landscape. An assemblage of ecosystems that are arranged in recognizable patterns and that exchange organisms and materials such as water.

landscape ecology. The study of dynamic interactions between the connected ecosystems forming a landscape and the environment, including human activities.

landscape mosaic. A patchwork of different components pieced together to form an overall landscape. The actual composition of the mosaic and the pattern in which the components are distributed will be unique to each landscape.

leach tube. A reusable cone-shaped plastic container used for growing seedlings. The containers can be mounted in a rack for easy transport and were named after inventor Ray Leach.

liner. A plant seedling grown in a long narrow tube (typically 10 inches tall and 1.5 inches in diameter) for convenient transplanting onto a revegetation site.

local ecological knowledge (LEK). Current and ever-expanding, useful knowledge about species and ecosystems, as gathered by people who live in rural landscapes in a sustainable manner. See also **traditional ecological knowledge (TEK)**.

macronutrient. A plant nutrient found at relatively high concentrations (> 500 mg kg-1) in plants. Usually refers to nitrogen, phosphorus, and potassium but may include calcium, magnesium, and sulfur.

microclimate. Ameliorated atmospheric conditions, relative to those of the macroclimate in the region, caused by community structure (e.g., shade, windbreaks) and processes (e.g., transpiration) in an ecosystem.

micronutrient. A plant nutrient found in relatively small amounts (< 100 mg kg-1) in plants. These are usually boron, chlorine, copper, iron, manganese, molybdenum, nickel, cobalt, and zinc.

mitigation. Mitigation includes (a) avoiding the impact altogether by not taking a certain action or parts of an action; (b) minimizing impacts by limiting the degree or magnitude of the action and its implementation; (c) rectifying the impact by repairing, rehabilitating, or restoring the affected environment; (d) reducing or eliminating the impact over time by preservation and maintenance operations during the life of the action; and (e) compensating for the impact by replacing or providing substitute resources or environments. The word *mitigation* is often used to refer to compensatory mitigation. See also **compensatory mitigation.**

mycorrhiza. A mutualistic symbiosis between plant and fungus localized in a root or rootlike structure in which energy moves primarily from plant to fungus and inorganic resources move from fungus to plant.

nurse plants. Plant species that protect or promote the growth of associated plants.

nutrient holding capacity. The ability of soil to absorb and retain nutrients so they will be available to the roots of plants.

organic matter present/content. The weight of decomposed plant and animal residue; expressed as a weight percentage of the soil material less than 2 millimeters in diameter.

passive restoration. Autonomous or autogenic recovery of a degraded ecosystem by means of the unassisted processes of resilience, succession, or natural regeneration.

performance standard. A value or threshold condition that is determined by monitoring and that, when attained, verifies that a particular objective has been achieved.

perturbation. An alteration of the function of a biological system, induced by external or internal mechanisms.

plant band. A long, narrow container for growing plants.

plant establishment period. The period of time following plant installation that is required to ensure success of the plantings installed at a restoration site without external support.

plant palette. The combination of plant species that are introduced to a restoration site. Also referred to as plant species palette or just species palette.

plugs. Herbaceous plants grown in small cylindrical or square containers that are longer than they are wide. The longer shape allows the plant to build root mass prior to transplanting. Also referred to as tubelings.

power auger. Motorized auger for boring shallow holes in the ground. Generally handheld by two workers.

prescribed fire or burning. The deliberate use of fire to manage a forest or some other type of natural area conducted for a single purpose or multiple purposes, including hazard reduction, control of understory vegetation, site preparation, disease control, and wildlife habitat enhancement as well as to kill targeted plant species or to favor the growth or presence of fire-dependent species of plants and animals.

prevegetated mats. Flat layers of vegetation (generally grasses, sedges, and forbs) grown in a soil medium incorporated into the top layer of a netting material made of polypropylene or plant fiber layers.

process. Dynamic aspect of an ecosystem or landscape, sometimes considered synonymous with function and including such interactions as transpiration, competition, parasitism, animal-mediated pollination and seed dispersal, mycorrhizal relationships, and other symbiotic relationships.

project requirements. A summarization of all of the site needs, stakeholder expectations, and imposed requirements for a restoration project.

project scope statement. A written statement describing the dimensions of a restoration project, including project goals and objectives, project requirements, the project budget, the project schedule, and any assumptions made by the project team.

propagule. Any plant reproductive structure, sexual and vegetative, such as a seed, spore, or rootstock, that proliferates.

pure live seed (PLS). The quantity of viable seed in harvested seed material derived by multiplying the germination and purity rates by the bulk quantity of seed.

purity rate. A measurement of bulk seed indicating the amount of seed and nonseed material.

reclamation. Conversion of land perceived as being relatively useless to a productive condition, commonly for agriculture and silviculture. Recovery of productivity is the main goal.

reference. One or more actual ecosystems (called reference sites), their written ecological descriptions, or information from secondary sources (e.g., historical photographs or accounts, paleoecological data) that serves as a basis for guiding the development of an ecological restoration project. See also **reference site(s)** and **reference model**.

reference model. An ecological description of an ecosystem that serves as a basis for preparing restoration plans; derived from the study of reference sites or from secondary sources of information.

reference site(s). One or more actual ecosystems on which restoration planning is based and that can serve as a basis for evaluating a completed restoration project.

rehabilitation. The recovery of ecosystem processes to regain normal function and ecosystem services without necessarily restoring the biodiversity of the reference or its projected trajectory.

resilience. The capacity of an ecosystem to tolerate or fully recover spontaneously from disturbance.

resistance. The capacity of an ecosystem to absorb the effects of disturbances with little or no change in structure and function.

resoiling. The process of artificially building or reconstructing a soil profile.

restoration ecology. The science on which the practice of ecological restoration is based and that provides the concepts and models on which practitioners depend. The science that advances the frontiers of theoretical ecology through studies of restored ecosystems and those that are undergoing restoration.

restrictive layer. A layer of earth that has one or more physical, chemical, or thermal properties that significantly reduce the movement of water and air through the soil. Restrictive layers limit plant growth by restricting the limits of the rooting zone.

revegetation. Establishment of plant cover on open land, usually with a limited number of species, irrespective of their provenance.

reverse backfilling. The refilling of a previously excavated area in reverse order from which the material was excavated (parent material, subsoil, topsoil).

rhizome. Stem of a plant that is capable of vegetative reproduction. Differing anatomically from true roots, rhizomes produce shoots above and roots below and are distinguished from a true root in possessing buds, nodes, and usually scalelike leaves.

salinization. Process by which soil comprising the root zone becomes increasingly more saline (salty) on account of the evaporation of irrigation water or another cause generally related to land use.

scarify. To break, scratch, or modify the surface of the soil. Also, to scratch the impervious seed coat of a hard seed.

seed bank. Location where seeds are stored for later purchase and use.

seed increase. A method of seed production using planted seedbeds to produce higher quantities of seeds in a controlled environment.

smothering. Technique for killing unwanted vegetation or preventing germination by covering the soil with some type of material (e.g., wood chips, weed control fabric, plastic) to exclude sunlight.

sod slabs. Blocks of sod removed from a wetland or native grassland or meadow containing the soil, the plant roots, and the aboveground vegetation.

soil electrical conductivity. The ability of soil to conduct electricity.

soil horizon. A layer of soil or soil material approximately parallel to the land surface and differing from adjacent genetically related layers in physical, chemical, and biological properties or characteristics, such as color, structure, texture, consistency, kinds and number of organisms present, and degree of acidity or alkalinity.

soil imprinting. A technique of using an angular tooth or foot (generally attached to a heavy roller) to create funnel-shaped depressions in the soil surface to promote plant growth. The technique differs from conventional methods of tillage (e.g., plowing, discing, cultivating, or drill seeding) in that it does not turn over the soil and entails minimal disruption of the surface litter. The depth of the depression is designed to permit sufficient water collection to allow germination of slower-growing plants.

soil inversion. A process of turning over the topsoil and bringing up the subsoil from three feet below the surface.

soil permeability. The ease with which gases, liquids, or plant roots penetrate or pass through a bulk mass of soil or a layer of soil.

soil ripping. The process of pulling a steel shank (tine or ripper) through soil to break up compacted subsurface layers. The shanks are more than forty-five centimeters long and spaced about the same distance apart. Also called subsoiling, deep ripping, or deep tillage.

soil seed bank. Viable seeds stored in the soil that are capable of germinating when appropriate conditions occur and that can replenish the vegetation after disturbance.

soil texture. The relative proportions of the various soil separates in a soil as described by the classes of soil texture (e.g., clay, clay loam, loam).

soil type. The lowest unit in the natural system of soil classification; a subdivision of a soil series and consisting of, or describing, soils that are alike in all characteristics, including the texture of the A horizon or plow layer. In terms of soil texture, soil type generally refers to the different sizes of mineral particles in a particular soil sample.

solarization. The use of solar energy to kill unwanted seeds and soil disease organisms; soil is tilled, irrigated, and covered with clear plastic.

species reintroduction. An attempt to reestablish a species in an area that was once part of its historical range but from which it has been extirpated.

stakeholder. A person or organization that is actively involved in the project or whose interests may be positively or negatively affected by execution or completion of the project.

state. The appearance, expression, or manifestation of an ecosystem or landscape as determined by species composition, the life-forms, sizes and abundance of individuals, and community structure.

stratification. The use of chemical and mechanical systems to break dormancy of seeds and promote germination.

structure. The physical appearance of a community as determined by the sizes, life-forms, abundance, and distribution of the predominant plant species. Also referred to as community structure.

swailing. Controlled burning for hazard reduction; also used for managing heath in Great Britain.

target. The intended long-term outcome (endpoint or goal) of a restoration project, which sometimes is not fully achieved until long after restoration project work has ceased.

target species. List of species (often special status species) for which habitat is being created. Restoration design is based on the combined habitat requirements of these species.

traditional cultural practices (TCPs). The application of traditional ecological knowledge that leads to the development and maintenance of cultural ecosystems.

traditional ecological knowledge (TEK). Ecological knowledge derived through societal experiences and perceptions that are accumulated within a traditional society through interaction with nature and natural resources. TEK commonly originates through trial and error and is frequently passed down to subsequent generations by oral tradition. See also **local ecological knowledge (LEK)**.

tree spade. A specialized machine consisting of a number of blades that encircle a tree, digging into the ground and then lifting the entire tree, including its roots and soil, out of the ground for relocation.

vegetation zonation. A vegetation pattern wherein certain assemblages of plant species occur in zones—often in bands adjacent to water bodies.

vegetative propagation. Propagation without pollination by way of separating vegetative parts (e.g., branches, stolons, rhizomes, buds) from the mother plant and planting them so they take root and grow.

viability. Determination of whether a seed is capable of germinating by establishing the presence of an embryo plant within the seed coat.

water budget. The determination of water needs for an area by totaling the various inputs from precipitation, groundwater, and other sources against the amount of water loss through runoff, transpiration, and ground infiltration.

water holding capacity. The ability of soils to hold water against the force of gravity and keep it available for use by most plants.

waterjet stinger. A high-pressure water pump plumbed to a long hollow pipe ("stinger"). The stinger is inserted into the ground, and the waterjet creates a long narrow hole. Used for planting woody cuttings along stream banks.

watershed. The line separating the waters flowing into different rivers or river basins; a narrow, elevated tract of ground between two drainage areas.

wattles. Long tubular rolls of plant material wrapped in twine or plastic netting used for erosion control and bank or slope stabilization. Wattles can be constructed from dormant stem cuttings or any other vegetative material (e.g., coir, rice straw, pine needles). The term *wattle* also refers to a fabrication of closely set posts interwoven with slender branches or reeds (wattle fences). See also **willow wattles**.

weed. Any undesired, uncultivated plant that grows in profusion so as to crowd out a desired crop or desired native vegetation.

weed-free straw. Harvested plant material that has been certified to be free from noxious weeds for use in straw bales, straw mulch, and straw wattles.

whips. Long woody cuttings.

willow wattles. Cylindrical bundles of live shrubby stems constructed from dormant willow cuttings; usually tied with twine or wire, varying in length and tapering at the ends. Used for erosion control and bank stabilization, dormant stem wattles can be constructed using any woody species that will root when in contact with moist soil. Also called fascines. See also **wattles**.

Adams, Lowell W., and Louise E. Dove. 1989. *Wildlife Reserves and Corridors in the Urban Environment*. Columbia, MD: National Institute for Urban Wildlife.

Allen, Michael F. 1991. *The Ecology of Mycorrhizae*. New York: Cambridge University Press.

Alwash, Suzanne. 2013. *Eden Again: Hope in the Marshes of Iraq*. Fullerton, CA: Tablet House Publishing.

Anderson, Bertin W., and Robert D. Ohmart. 1985. "Riparian Revegetation as a Mitigating Process in Stream and River Restoration." In *The Restoration of Rivers and Streams*, ed. James A. Gore, 41–79. Boston: Butterworth Publishers.

Anderson, M. Kat. 2001. "The Contribution of Ethnobiology to the Reconstruction and Restoration of Historic Ecosystems." In *The Historical Ecology Handbook: A Restorationist's Guide to Reference Ecosystems*, ed. Dave Egan and Evelyn A. Howell, 55–72. Washington, DC: Island Press.

———. 2005. "Tending the Wild: Native American Knowledge and the Management of California's Natural Resources." Berkeley: University of California Press.

Aqua Dam Inc. 2012. Website. http://www.aquadam.com.

Bainbridge, David A. 2007. *A Guide for Desert and Dryland Restoration*. Washington, DC: Island Press.

Baird, Kathryn, and John Rieger. 1989. "A Restoration Design for Least Bell's Vireo Habitat in San Diego County." In *Proceedings of the California Riparian Systems Conference: Protection, Management, and Restoration for the 1990's*, ed. D. L. Abell, 462–67. Berkeley, CA: Pacific Southwest Forest and Range Experiment Station, Forest Service, US Department of Agriculture.

Blackburn, Thomas C., and Kat Anderson, eds. 1993. *Before the Wilderness: Environmental Management by Native Californians*. Menlo Park, CA: Ballena.

Bonham, Charles D. 1989. *Measurements of Terrestrial Vegetation*. New York: Wiley-Interscience.

Boustany, Ronald G. 2003. "A Pre-vegetated Mat Technique for the Restoration of Submerged Aquatic Vegetation." *Ecological Restoration* 21:87–94.

Bradley, Joan. 2002. *Bringing Back the Bush: The Bradley Method of Bush Regeneration*. Sydney, Australia: Reed New Holland.

Bradshaw, Anthony D., and M. J. Chadwick. 1980. *The Restoration of Land: The Ecology and Reclamation of Derelict and Degraded Land*. Berkeley: University of California Press.

Burkhart, Brad. 2006. "Selecting the Right Container for Revegetation Success with Tap-Rooted and Deep-Rooted Chaparral and Oak Species." *Ecological Restoration* 24:87–92.

California Tahoe Conservancy. 2012a. "Trout Creek Restoration." http://tahoe.ca.gov/trout-creek-restoration-69.aspx.

California Tahoe Conservancy. 2012b. "Upper Truckee Marsh." http://tahoe.ca.gov/upper-truckee-marsh-69.aspx.

Clewell, Andre, John Rieger, and John Munro. 2000. *Guidelines for Developing and Managing Ecological Restoration Projects.* 1st ed. Washington, DC: Society for Ecological Restoration.

———. 2005. *Guidelines for Developing and Managing Ecological Restoration Projects.* 2nd ed. Washington, DC: Society for Ecological Restoration. http://www.ser.org.

Clewell, Andre F. 1999. "Restoration of Riverine Forest at Hall Branch of Phosphate-Mined Land, Florida." *Restoration Ecology* 7:1–14.

Clewell, Andre F., and James Aronson. 2013. *Ecological Restoration: Principles, Values, and Structure of an Emerging Profession.* 2nd ed. Washington, DC: Island Press.

Clewell, Andre F., and James Aronson. 2007. "Reference Models and Developmental Trajectories." In Andre F. Clewell and James Aronson, *Ecological Restoration: Principles, Values, and Structure of an Emerging Profession,* 75–87. Washington, DC: Island Press.

Cox, George. 1999. *Alien Species in North America and Hawaii: Impacts on Natural Ecosystems.* Washington, DC: Island Press.

Daigle, Jean-Marc, and Donna Havinga. 1996. *Restoring Nature's Place.* Schomberg, Ontario: Ecological Outlook Consulting and Ontario Parks Association.

Diamond, Jared. 1975. "The Island Dilemma Lessons of Modern Biogeographic Studies for the Design of Natural Reserves." *Biological Conservation* 7:129–46.

Dorner, J. 2002. "An Introduction to Using Native Plants in Restoration Projects." Center for Urban Horticulture, University of Washington; USDI Bureau of Land Management; US Environmental Protection Agency.

Doyle, Michael, and David Straus. 1976. *How to Make Meetings Work.* New York: Berkley Publishing Group.

Edmonds, Michael. 2001. "The Pleasures and Pitfalls of Written Records." In *The Historical Ecology Handbook: A Restorationist's Guide to Reference Ecosystems,* ed. Dave Egan and Evelyn A. Howell, 73–99. Washington, DC: Island Press.

Egan, Dave, and Evelyn A. Howell. 2001. "The Historical Ecology Handbook: A Restorationist's Guide to Reference Ecosystems." In *The Historical Ecology Handbook: A Restorationist's Guide to Reference Ecosystems,* ed. Dave Egan and Evelyn A. Howell, 1–23. Washington, DC: Island Press.

Elzinga, Caryl L., Daniel W. Salzer, and John W. Willoughby. 1998. *Measuring and Monitoring Plant Populations.* Denver, CO: Bureau of Land Management, National Business Center.

Elzinga, Caryl L., Daniel W. Salzer, John W. Willoughby, and James P. Gibbs. 2001. *Monitoring Plant and Animal Populations.* Malden, MA: Blackwell Science.

Falk, Donald A., Margaret A. Palmer, and Joy B. Zedler. 2006. *Foundations of Restoration Ecology.* Washington, DC: Island Press.

Fogerty, James E. 2001. "Oral History: A Guide to Its Creation and Use." In *The Historical Ecology Handbook: A Restorationist's Guide to Reference Ecosystems,* ed. Dave Egan and Evelyn A. Howell, 101–20. Washington, DC: Island Press.

Frame, J. Davidson. 1995. *Managing Projects in Organizations: How to Make the Best Use of Time, Techniques, and People.* San Francisco: Jossey-Bass.

Friederici, Peter, ed. 2003. *Ecological Restoration of Southwestern Ponderosa Pine Forests.* Washington, DC: Island Press.

Gray, Donald H., and Robbin B. Sotir. 1996. *Biotechnical and Soil Bioengineering Slope Stabilization: A Practical Guide for Erosion Control.* New York: John Wiley & Sons.

Griggs, F. Thomas. 2009. *California Riparian Habitat Restoration Handbook.* Chico, CA: River Partners.

Grossinger, Robin. 2001. "Documenting Local Landscape Change: The San Francisco Bay Area Historical Ecology Project." In *The Historical Ecology Handbook: A Restorationist's Guide to Reference Ecosystems*, ed. Dave Egan and Evelyn A. Howell, 425–39. Washington, DC: Island Press.

Hall, Jason, Michael Pollock, and Shirley Hob. 2011. "Methods for Successful Establishment of Cottonwood and Willow along an Incised Stream in Semiarid Eastern Oregon, USA." *Ecological Restoration* 29:261–69.

Hammer, Donald A. 1997. *Creating Freshwater Wetlands.* Boca Raton, FL: Lewis.

Hammer, Joshua. 2006. "Return to the Marsh: The Effort to Restore the Marsh Arabs' Traditional Way of Life in Southern Iraq—Virtually Eradicated by Saddam Hussein—Faces New Threats." http://www.smithsonianmag.com/people-places/return-to-the-marsh-132043707.

Harris, James A., Paul Birch, and John Palmer. 1996. *Land Restoration and Reclamation: Principles and Practice.* Singapore: Addison Wesley Longman.

Harris, Jim. 2009. "Perspective: Soil Microbial Communities and Restoration Ecology: Facilitators or Followers?" *Science* 325:573–74.

Hindle, Tim. 1998. *Managing Meetings.* New York: DK Publishing.

Hoag, Chris, and Jon Fripp. 2002. *Streambank Soil Bioengineering Field Guide for Low Precipitation Areas.* Fort Worth, TX: USDA-NRCS Aberdeen Idaho Plant Materials Center and NRCS National Design, Construction, and Soil Mechanics Center.

Hoag, J. Chris, and Dan Ogle. 2008. "The Stinger: A Tool to Plant Unrooted Hardwood Cuttings. Plant Materials Technical Note No. 6." http://chapter.ser.org/northwest/files/2012/08/NRCS_TN6_the_stinger.pdf.

Hoag, John C., Boyd Simonson, Brent Cornforth, and Loren St. John. 2001. "Waterjet Stinger: A Tool to Plant Dormant Unrooted Cuttings of Willow, Cottonwood, Dogwood and Other Species." http://www.plant-materials.nrcs.usda.gov/pubs/idpmctn1083.pdf.

Howell, Evelyn A., John A. Harrington, and Stephen B. Glass. 2012. *Introduction to Restoration Ecology.* Washington, DC: Island Press.

IUCN WCPA Ecological Restoration Taskforce. 2012. *Ecological Restoration for Protected Areas: Principles, Guidelines and Best Practices.* Gland, Switzerland: International Union for Conservation of Nature and Natural Resources.

Karr, James R., and Ellen W. Chu. 1999. *Restoring Life in Running Waters: Better Biological Monitoring.* Washington, DC: Island Press.

Katan, Jaacov, and James E. DeVay. 1991. *Soil Solarization.* Boca Raton, FL: CRC Press.

Kentula, Mary E., Robert P. Brooks, Stephanie E. Gwin, Cindy C. Holland, Arthur D. Sherman, and Jean C. Sifneos. 1993. *An Approach to Improving Decision Making in Wetland Restoration and Creation*. Boca Raton, FL: C.K. Smoley.

King County Department of Natural Resources. 2011. "Get Involved in the Native Plant Salvage Program." http://www.kingcounty.gov/environment/stewardship/volunteer/plant-salvage-program.aspx.

Kloetzel, S. 2004. "Revegetation and Restoration Planting Tools: An In the Field Perspective." *Native Plants* 5 (1):34–42.

Kondolf, G. Mathias. 1995. "Five Elements for Effective Evaluation of Stream Restoration." *Restoration Ecology* 3:133–36.

Kotler, Philip. 1999. *Marketing Management: Millennium Edition*. Upper Saddle River, NJ: Prentice-Hall.

Krebs, Charles J. 1989. *Ecological Methodology*. New York: Harper Collins.

Lambrecht, Susan C., and Antonia D'Amore. 2010. "Solarization for Non-native Plant Control in Cool, Coastal California." *Ecological Restoration* 28:424–25.

Larson, Marit. 1995. "Developments in River and Stream Restoration in Germany." *Restoration and Management Notes* 13:77–83.

Leck, Mary A., V. Thomas Parker, and Robert L. Simpson, eds. 1989. *Ecology of Soil Seed Banks*. San Diego, CA: Academic Press.

Lewis, Henry T. 1993. "Patterns of Indian Burning in California: Ecology and Ethnohistory." In Thomas C. Blackburn and Kat Anderson, *Before the Wilderness: Environmental Management by Native Californians*, 55–116. Menlo Park, CA: Ballena.

Luce, Charles H. 1997. "Effectiveness of Road Ripping in Restoring Infiltration Capacity of Forest Roads." *Restoration Ecology* 5:265–70.

Maehr, David S., Thomas S. Hoctor, and Larry D. Harris. 2001. "The Florida Panther: A Flagship for Regional Restoration." In *Large Mammal Restoration: Ecological and Sociological Challenges in the 21st Century*, ed. David S. Maehr, Reed F. Noss, and Jeffery L. Larkin, 293–312. Washington, DC: Island Press.

Manley, Patricia N., Beatrice Van Horne, Julie K. Roth, William J. Zielinski, Michelle M. McKenzie, Theodore J. Weller, Floyd W. Weckerly, and Christina Vojta. 2006. *Multiple Species Inventory and Monitoring Technical Guide*. General Technical Report WO-73. Washington, DC: US Department of Agriculture, Forest Service.

March, Rosaleen G., and Elizabeth H. Smith. 2011. "Combining Available Spatial Data to Define Restoration Goals." *Ecological Restoration* 29:252–60.

Margoluis, Richard, and Nick Salafsky. 1998. *Measures of Success: Designing, Managing and Monitoring Conservation and Development Projects*. Washington, DC: Island Press.

Martin, Paula, and Karen Tate. 1997. *Project Management: Memory Jogger*. Methuen, MA: GOAL/QPC.

McHarg, Ian L. 1969. *Design with Nature*. New York: American Museum of Natural History/ Natural History Press.

Merriam-Webster Inc. 2003. *Merriam-Webster's Collegiate Dictionary*. Springfield, MA: Merriam Webster Inc.

Middleton, Beth. 1999. *Wetland Restoration: Flood Pulsing and Disturbance Dynamics*. New York: John Wiley & Sons.

Montalvo, Arlee M., and Norman C. Ellstrand. 2000. "Transplantation of the Subshrub *Lotus scoparius*: Testing the Home-Site Advantage Hypothesis." *Conservation Biology* 14:1034–45.

Montalvo, Arlee M., Paul A. McMillan, and Edith B. Allen. 2002. "The Relative Importance of Seeding Method, Soil Ripping and Soil Variables on Seeding Success." *Restoration Ecology* 10:52–67.

Moore, Charles W. 1960. "Hadrian's Villa." *Perspecta: The Yale Architectural Journal* 6:16–27.

Morrison, Michael L. 2009. *Restoring Wildlife: Ecological Concepts and Practical Applications*. Washington, DC: Island Press.

Morrison, Michael L., Thomas A. Scott, and Tracy Tennant. 1994. "Wildlife-Habitat Restoration in an Urban Park in Southern California." *Restoration Ecology* 2:17–30.

Muhar, Susanne, Stefan Schmutz, and Mathias Jungwirth. 1995. "River Restoration Concepts: Goals and Perspectives." *Hydrobiologia* 303:183–94.

Murphy, Stephen D., Jay Flanagan, Kevin Noll, Dana Wilson, and Bruce Duncan. 2007. "How Incomplete Exotic Species Management Can Make Matters Worse: Experiments in Forest Restoration in Ontario, Canada." *Ecological Restoration* 25:85–93.

Naveh, Zev. 1989. "Neot Kedumim." *Restoration and Management Notes* 5:9–13.

Naveh, Zev, and Arthur S. Lieberman. 1984. *Landscape Ecology: Theory and Application*. Berlin, Germany: Springer-Verlag.

Nelson, Harold L. 1987. "Prairie Restoration in the Chicago Area." *Restoration and Management Notes* 5:60–67.

Newton, Adrian, and Philip Ashmole, eds. 2010. "Carrifran Wildwood Project: Native Woodland Restoration in the Southern Uplands of Scotland," Management plan. Wildwood Group of the Borders Forest Trust.

Oldham, Jon A. 1989. "The Hydrodriller: An Efficient Technique for Installing Woody Stem Cuttings." In *Proceedings of a Symposium: First Annual Meeting of the Society for Ecological Restoration*, ed. Glenn Hughes and Tom Bonnicksen, 69–78. Madison, WI: Society for Ecological Restoration.

Packard, Stephen, and Cornelia F. Mutel, eds. 1997. *The Tallgrass Restoration Handbook: For Prairies, Savannas, and Woodlands*. Washington, DC: Island Press.

Pierce, Gary J. 1993. *Planning Hydrology for Constructed Wetlands*. Poolesville, MD: Wetland Training Institute.

PMBOK. 2008. *A Guide to the Project Management Body of Knowledge (PMBOK Guide)*. 4th ed. Newtown, PA: Project Management Institute.

Pritchard, Carl L., ed. 1997. *Risk Management: Concepts and Guidance*. Arlington, VA: ESI International.

Rea, Amadeo M. 1983. *Once a River: Bird Life and Habitat Changes in the Middle Gila*. Tucson: University of Arizona Press.

Savory, Allan. 1998. *Holistic Management: A New Framework for Decision Making*. Washington, DC: Island Press.

Smith, Daniel S., and Paul C. Hellmund. 2006. *Designing Greenways: Sustainable Landscapes for Nature and People*. Washington, DC: Island Press.

Southwood, Thomas R. E., and Peter A. Henderson. 2000. *Ecological Methods*. Oxford: Blackwell Science.

St. John, Loren, Brent Cornforth, Boyd Simonson, Dan Ogle, and Derek Tilly. 2008. *Calibrating the Truax Rough Rider Seed Drill for Restoration Plantings*. Plant Materials Technical Note No. 20. Boise, ID: USDA Natural Resources Conservation Service.

Schiechtl, Hugo M., and Roland Stern. 1996. *Ground Bioengineering Techniques for Slope Protection and Erosion Control*. Oxford: Blackwell.

———. 1997. *Water Bioengineering Techniques for Watercourse Bank and Shoreline Protection*. Oxford: Blackwell.

Sutherland, William J., ed. 1996. *Ecological Census Techniques: A Handbook*. Cambridge: Cambridge University Press.

Tongway, David J. 2010. "Teaching the Assessment of Landscape Function in the Field: Enabling the Design and Selection of Appropriate Restoration Techniques." *Ecological Restoration* 28:182–87.

Tongway, David J., and John A. Ludwig. 2011. *Restoring Disturbed Landscapes: Putting Principles into Practice*. Washington, DC: Island Press.

Valentin, Anke, and Joachim H. Spangenberg. 2000. "A Guide to Community Sustainability Indicators." *Environmental Impact Assessment Review* 20:381–92.

van Andel, Jelte, and James Aronson. 2012. *Restoration Ecology: The New Frontier*. Oxford: Blackwell.

Weekley, Carl W., Eric S. Menges, Dawn Berry-Greenlee, Marcia A. Rickey, Gretel L. Clarke, and Stacy A. Smith. 2011. "Burning More Effective than Mowing in Restoring Florida Scrub." *Ecological Restoration* 29:357–73.

White, Peter S., and Joan L. Walker. 1997. "Approximating Nature's Variation: Selecting and Using Reference Information in Restoration Ecology." *Restoration Ecology* 5:338–49.

Zentner, John. 1994. "Enhancement, Restoration and Creation of Freshwater Wetlands." In *Applied Wetlands Science and Technology*, ed. Donald M. Kent, 127–66. Boca Raton, FL: Lewis.

John Rieger

John Rieger has spent the past thirty years of his professional career involved with various aspects of ecological restoration. As cofounder of the Society for Ecological Restoration (SER) in 1987, John became its first president in 1988. John has worked to promote ecological restoration, presenting workshops throughout the United States and in Canada and England. He was presented the SER Service Award in 1995 and the Golden Trowel Award in 1997.

Educated as a wildlife biologist at San Diego State University, John became a Certified Wildlife Biologist in 1978 from The Wildlife Society. John was hired as the first district biologist at Caltrans in San Diego in 1979. He performed some of the pioneering work on endangered Least Bell's (Vireo bellii pusillus) habitat restoration, working with his fellow biologists and landscape architects at Caltrans. John and Ray Traynor led the team that received the SER 1996 Model Project of the Year Award. John achieved a Project Management Professional designation from the Project Management Institute in 2004. He was a senior biologist and project manager for Caltrans.

John continues his involvement with ecological restoration by volunteering on committees for SER, conducting ecological restoration workshops, and managing his consulting firm, which he founded in 1984 with his wife, Mary F. Platter-Rieger.

John Stanley

John Stanley is a restoration ecologist with WWWRESTORATION, a consulting firm specializing in the protection, restoration, and management of watersheds, waterways, and wetlands. John Stanley has more than forty years of academic, agency, nonprofit, and consulting experience in environmental planning, biotic impact analysis, habitat mitigation, soil conservation, ecological restoration, and watershed management.

Formerly a principal and licensed landscape contractor with The Habitat Restoration Group, John was responsible for the planning, design, construction/installation, maintenance, and monitoring of land reclamation, habitat mitigation, and ecological restoration projects throughout northern and central California.

John is a founding member of the Society for Ecological Restoration (SER). He recently completed serving as SER's US western regional representative and is continuing to serve as the chair of SER's ad hoc committee for Ecological Restoration Practitioner Certification. He lives in Paradise, California.

Ray Traynor

Ray Traynor is a member of the executive team at the San Diego Association of Governments, where he is principal regional planner. Ray is a California Registered Landscape Architect. He lives in San Diego with his wife, Katrin, where they enjoy hiking and tandem cycling. Ray received his bachelor's degree in landscape architecture from the School of Architecture and Environmental Design at Cal Poly San Luis Obispo in 1983. In 1997, Ray obtained an MBA from San Diego State University, with an emphasis on entrepreneurial management.

Ray's interest in ecological restoration began in the mid 1980s when he worked as a landscape architect for the California Department of Transportation. Ray's thirty-year career has included work in landscape architecture, ecological restoration, project management, business development, product management, mobile applications development, and transportation planning.

Wildlife Restoration: Techniques for Habitat Analysis and Animal Monitoring, by Michael L. Morrison

Ecological Restoration of Southwestern Ponderosa Pine Forests, edited by Peter Friederici, Ecological Restoration Institute at Northern Arizona University

Ex Situ Plant Conservation: Supporting Species Survival in the Wild, edited by Edward O. Guerrant Jr., Kayri Havens, and Mike Maunder

Great Basin Riparian Ecosystems: Ecology, Management, and Restoration, edited by Jeanne C. Chambers and Jerry R. Miller

Assembly Rules and Restoration Ecology: Bridging the Gap Between Theory and Practice, edited by Vicky M. Temperton, Richard J. Hobbs, Tim Nuttle, and Stefan Halle

The Tallgrass Restoration Handbook: For Prairies, Savannas, and Woodlands, edited by Stephen Packard and Cornelia F. Mutel

The Historical Ecology Handbook: A Restorationist's Guide to Reference Ecosystems, edited by Dave Egan and Evelyn A. Howell

Foundations of Restoration Ecology, edited by Donald A. Falk, Margaret A. Palmer, and Joy B. Zedler

Restoring the Pacific Northwest: The Art and Science of Ecological Restoration in Cascadia, edited by Dean Apostol and Marcia Sinclair

A Guide for Desert and Dryland Restoration: New Hope for Arid Lands, by David A. Bainbridge
Restoring Natural Capital: Science, Business, and Practice, edited by James Aronson, Suzanne J. Milton, and James N. Blignaut

Old Fields: Dynamics and Restoration of Abandoned Farmland, edited by Viki A. Cramer and Richard J. Hobbs

Ecological Restoration: Principles, Values, and Structure of an Emerging Profession, by Andre F. Clewell and James Aronson

River Futures: An Integrative Scientific Approach to River Repair, edited by Gary J. Brierley and Kirstie A. Fryirs

Large-Scale Ecosystem Restoration: Five Case Studies from the United States, edited by Mary Doyle and Cynthia A. Drew

New Models for Ecosystem Dynamics and Restoration, edited by Richard J. Hobbs and Katharine N. Suding

Cork Oak Woodlands in Transition: Ecology, Adaptive Management, and Restoration of an Ancient Mediterranean Ecosystem, edited by James Aronson, João S. Pereira, and Juli G. Pausas

Restoring Wildlife: Ecological Concepts and Practical Applications, by Michael L. Morrison

Restoring Ecological Health to Your Land, by Steven I. Apfelbaum and Alan W. Haney

Restoring Disturbed Landscapes: Putting Principles into Practice, by David J. Tongway and John A. Ludwig

Intelligent Tinkering: Bridging the Gap between Science and Practice, by Robert J. Cabin

Making Nature Whole: A History of Ecological Restoration, by William R. Jordan and George M. Lubick

Human Dimensions of Ecological Restoration: Integrating Science, Nature, and Culture, edited by Dave Egan, Evan E. Hjerpe, and Jesse Abrams

Plant Reintroduction in a Changing Climate, edited by Joyce Maschinski and Kristin E. Haskins

Tidal Marsh Restoration: A Synthesis of Science and Management, edited by Charles T. Roman and David M. Burdick

Ecological Restoration: Principles, Values, and Structure of an Emerging Profession, 2nd ed., Andre F. Clewell and James Aronson